LeadingGreen

LeadingGreen: LEED® Green Associate Exam v4 Study Guide brings together a concise introduction to the core concepts behind the LEED green building rating system with efficient, effective test preparation for the LEED® Green Associate Exam. The guide provides an essential foundation in principles of sustainable design and construction for students and professionals in the areas of architecture, engineering, construction, real estate development, urban planning, and environmental policy.

If you are interested in passing the LEED® Green Associate Exam, this is the right place to start. The background, objectives, and evolution of green building standards – as well as what actually goes into a LEED building – are covered in a clear and straightforward manner. Tailored to precise exam expectations, the guide breaks down the key LEED categories one credit at a time.

Drawing upon the author's extensive experience as a LEED educator for universities and professional firms around the world, the guide includes proven tips and tricks that have helped over 10,000 people pass the Green Associate Exam.

An ideal resource for self-study, *LeadingGreen: LEED® Green Associate Exam v4 Study Guide* will benefit readers from all built environment professions in which LEED credentials are an asset.

Lorne Mlotek, BASc, LEED AP BD+C, O+, is the President/CEO at LeadingGreen Training and Consulting Inc. A civil engineer with over 20 years of experience in sustainability initiatives, Mlotek has taught LEED certification courses internationally in architecture, engineering, and construction companies and in partnership with 250 universities.

LeadingGreen

LEED® Green Associate Exam v4
Study Guide

Lorne Mlotek

Routledge
Taylor & Francis Group

NEW YORK AND LONDON

Designed cover image: Photograph by Danist Soh on Unsplash

First published 2023
by Routledge
605 Third Avenue, New York, NY 10158

and by Routledge
4 Park Square, Milton Park, Abingdon, Oxon, OX14 4RN

Routledge is an imprint of the Taylor & Francis Group, an informa business

ISBN: 978-1-032-52274-6 (hbk)
ISBN: 978-1-032-48557-7 (pbk)
ISBN: 978-1-003-40585-6 (ebk)

DOI: 10.1201/9781003405856

Typeset in Times New Roman
by codeMantra

Access the Support Material: www.routledge.com/9781032485577

Contents

Introduction   vi
Author Biography   viii

1 Green Buildings and Sustainability   1

2 The Governing Bodies of LEED – USGBC + GBCI   9

3 Integrative Process (1 Point)   33

4 Location and Transportation (LT)   39

5 Sustainable Sites (SS)   53

6 Water Efficiency   69

7 Energy and Atmosphere   80

8 Materials and Resources   99

9 Indoor Environmental Quality   115

10 Innovation in Design   131

11 Regional Priority   135

Appendix I: Additional Reading   137
Appendix II: LEED Green Associate Exam Test Structure   138
Appendix III: Glossary   139
References   143
Index   145

Introduction

This textbook is an introduction to the basics of green building and a re-source prepared with the intention of helping you pass the LEED® Green Associate Exam. LEED, an acronym for Leadership in Energy and Environmental Design, is the premier rating system employed in the building industry. It is aimed at appraising and assessing the performance of buildings, thereby improving building efficiency and making them more sustainable. This handbook presents concepts and strategies for green building design adopting best practices that are both environmentally sound and resource-efficient. It is logically split into distinct chapters beginning with a comprehensive introduction and briefing on green buildings and the governing bodies of LEED while also highlighting exactly how the LEED rating system evaluates and gauges a building's ability to sustain. This guide has been penned keeping in mind all reading levels regardless of their prior background knowledge or experience.

Green building evaluations are carried out in conformity with core areas, reflecting their impact on people (social benefits), the environment (environmental benefits), and profit or financial gains (economic benefits). This is popularly referred to as the triple bottom line. The workbook explores how LEED is capable of incorporating existing standards and taking advantage of them in the rating system for the general interest and benefit of the population.

At the end of each chapter, you will find realistic practice questions and explanations for your studies. Following completion of this textbook, you will be in a position to confidently address and discuss sustainability as it pertains to the built environment while understanding and perceiving the key concepts tested on the LEED Green Associate Exam.

Each chapter follows a similar structure and rhythm, and it helps:

1   Understand the problem and traditional industry processes
2   Learn how LEED addresses and tackles the issue
3   Figure out what LEED requires to comply with the rating system

From my personal standpoint, those who genuinely understand the functioning of the overall LEED rating system and are familiar with how each credit/strategy interacts with one another are those who have the easiest time and fare well on their exam. Bear this in mind as you brace yourself for your exam and ensure you do not get lost in the details as the LEED Green Associate Exam tests your high-level understanding and knowledge of LEED and green buildings.

Author Biography

Located in the heart of Toronto, Canada, LeadingGreen Training and Consulting Inc. is a leading education and service provider with more than a decade of long-standing experience in the sustainable design and construction industry. LeadingGreen has a highly competent professional team of experts who have experience consulting on projects and training across the globe, including the United Kingdom, the Middle East, and all throughout North America. Right from LEED® accreditation training to LEED project management, commissioning, energy modeling, and auditing, their core services also encompass energy retrofits for existing buildings.

LeadingGreen has prepared over 10,000 people for their LEED Green Associate and Accredited Professional exams, fruitfully achieving an exceptionally high passing rate. LeadingGreen's team has collective experience in designing, constructing, and renovating over 50 LEED-registered and certified commercial, institutional, and high-rise residential buildings on a global scale. They offer an extensive range of sustainable design and construction services to meet and satisfy their client's green building objectives, in conjunction with energy analysis, renewable energy integration, life cycle costing, energy auditing, and ongoing measurement and verification services. LeadingGreen's aptitude, proficiency, and knowledge in the cost-effective and energy-efficient design ensure the client's objectives are always fulfilled through a holistic project.

Lorne Mlotek, LeadingGreen's President and lead instructor, is personally responsible for creating and conducting over 500 workshops in favor of the academic, government, and private clients. Lorne graduated with honors from Civil Engineering at the University of Toronto and is currently a LEED Accredited Professional (AP+) specializing in Building Design and Construction (BD+C) and Operations and Maintenance (O+M). He has been personally vetted by the USGBC and recognized as an esteemed member of the USGBC faculty for his past work and current professional training courses.

1 Green Buildings and Sustainability

According to the U.S. Environmental Protection Agency (EPA), sustainability is defined as follows.

> Everything that we need for our survival and well-being depends, either directly or indirectly, on our natural environment. To pursue sustainability is to create and maintain the conditions under which humans and nature can exist in productive harmony to support present and future generations.[1]

Or *anecdotally,* the mindset is that we do not inherit the Earth from our ancestors but rather borrow it from our children.

Presently the construction and operation of buildings are clearly responsible for a myriad of detrimental effects on planet Earth. They account for one of the largest consumers of energy and electricity, and they, directly and indirectly, emit an immense amount of greenhouse gases, thereby interfering with the ecological footprint and contributing to climate change.

> The greenhouse effect is the way in which heat is trapped close to Earth's surface by "greenhouse gases." These heat-trapping gases can be thought of as a blanket wrapped around Earth, keeping the planet toastier than it would be without them.[2]

DOI: 10.1201/9781003405856-1

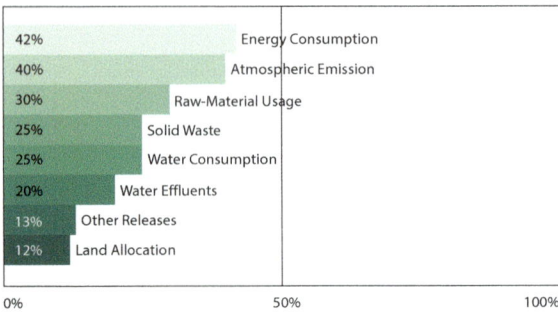

Figure 1.1 Graph to show the percentages of our energy consumption, emissions, waste and land allocation related to building use.

Owing to this phenomenon being responsible for the artificial warming of our planet, it results in the deterioration of ecosystems.

Consequently, the planet now experiences frequent and severe variations in weather conditions like warmer summers, colder winters, more vicious weather events or climatic aberrations, and even drought. This textbook has been compiled assuming the reader truly realizes and understands the adverse repercussions of climate fluctuations through the emission of greenhouse gases primarily triggered by generating energy to power buildings. We will explore alternative approaches wherein LEED® attempts to combat detrimental human interactions on our climate. Although the climate crisis is still a critical issue debated to this day, one fact continues to be the case: The Earth possesses an insignificant number of resources that we utilize to fuel our daily lives. Regardless of climate change, the human consumption rate of the Earth or land resources is exorbitantly high. It is often observed that as our demand for non-renewable resources such as gas, oil, and coal radically escalates, their supply continues to shrink, which will persist in increasing the price of energy drastically. Eventually, in due course, energy will be inaccessible and unaffordable as our short-sighted consumption rate is far too high to sustain our needs (Figure 1.1).

Energy Production and Resources

The world population rapidly continues to rise and climb at an exponential rate. In the wake of the population gaining access to electricity and energy, drastic changes will be required to be made to address and meet our future energy requirements. In this situation, there exists only two means equipped to satisfy our ever-increasing growing demand for energy:

1   Boost the production of energy through traditional and renewable facilities.
2   Consume a lesser amount of energy than what we currently use.

Source   **Use**

3 times more expensive to Produce than Reuse

Figure 1.2 To satisfy our growing demand for energy, we must either increase production of energy through traditional and renewable facilities or reduce
the amount of energy we currently use.

While both alternatives are viable, under the current circumstances, only
one is financially favorable at the moment. Irrespective of the form of energy
source we decide on singling out and increasing, be it nuclear, fossil fuels,
or renewables, it is more expensive to generate a new KWH (kilowatt hour)
of energy versus reducing where we currently consume that same KWH of
energy. Economically, it makes perfect sense to maximize all possible means
of reducing our consumption rather than focusing on creating new production facilities. In view of the fact that buildings are the largest consumers of
energy and electricity, it only seems logical to commence our path of reduction with the built environment (Figure 1.2).

The NEED for LEED

Sustainable constructions or green buildings emerged in response to limiting and restricting the damaging and detrimental repercussions of the
building industry on the environment, economy, and society. The built environment refers to the man-made surroundings that encompass buildings
and transportation systems that form neighborhoods, which leads to the
creation of the urban system. It is the largest consumer of energy and producer of greenhouse gases.

The primary goal of LEED is to eliminate greenwashing, the act or practice of making a product, policy, activity, etc., appear to be more eco-friendly
than reality.[3] Prior to LEED, it was possible for anyone to theoretically
construct a building without any sustainable attributes, paint it green and
market it as a "green building." LEED fills this void and bridges the gap, of-
fering the public as well as potential investors a level of trust that the building positively impacts the economy, environment, and society.

Traditional Practices vs. Sustainable Practices

Traditionally, the building industry operates in a vacuum and only evaluates the present time. Stakeholders throughout the design and construction industry work alone and focus solely on their project goals. On the contrary, sustainable practices are cohesive, ensuring all project stakeholders are in sync while being aware of their potential impact on the project's overall goal, both now in the present, as well as in the future (Figure 1.3).

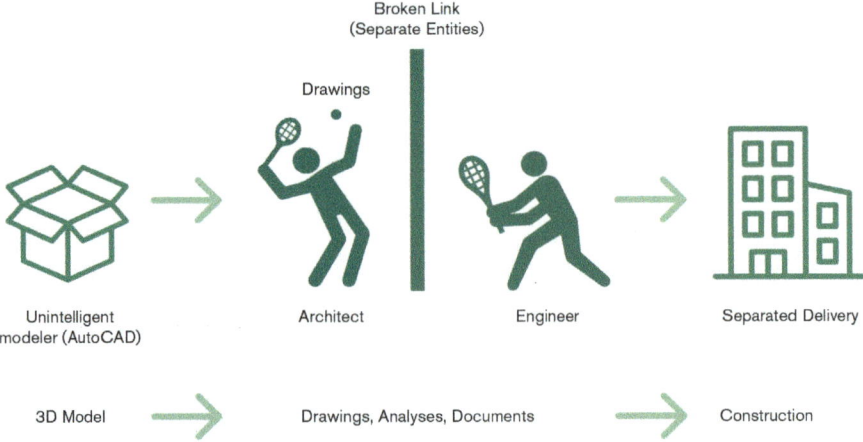

Figure 1.3 The stages and entities involved in designing and implementing a new sustainable practice.

For example:

Traditional Practices	Sustainable Practices
During the site selection phase, only a project's budget shall be taken into account, and the environmental impact of clearing land for development is completely overlooked and ignored. Land development poses a grave threat to wildlife and habitat destruction, burdening our natural resource supply and requiring more services, parking lots, and transportation systems to be built. Conventional site selection does not pay heed to transit accessibility to the masses, shared parking, existing services, or the location of prime farmlands.	Site selection analysis is resorted to carefully screen the ideal location or site, considering factors such as existing climate, water, soil, and habitat. Likewise, community connectivity, accessibility to mass transit, opportunities to share parking, and avoiding building on sensitive land are also determinants that are duly addressed.

Traditional Practices	*Sustainable Practices*
No preventive and cautionary measures are undertaken to limit potential problems such as air pollution, water pollution, greenhouse gases, and the release of toxic chemicals into the environment during construction.	Sustainable construction activities and ongoing operations explore waste management and pollution control practices and strategies.
Best practices in water and energy consumption are neglected.	Designing and implementing energy and water-efficient systems as well as monitoring the system to conserve energy for the future.
Under the traditional approach, initial costs are the only relevant cost of the building.	Operation and maintenance costs are marked down owing to the deployment of more efficient utilities and premium-quality buildings.

Triple Bottom Line

The Social, Environmental, and Economic Benefits of Green buildings are referred to as the Triple Bottom Line (Figure 1.4). Green building evaluation is an appraisal function based on their impact exerted on People (Social benefits), Planet (Environmental benefits), and Profit (Economic benefits). For that very reason, the triple bottom line is also known as the 3 Ps.

All LEED-certified projects achieve these three sustainability goals, which is how they earn the status of a green building.

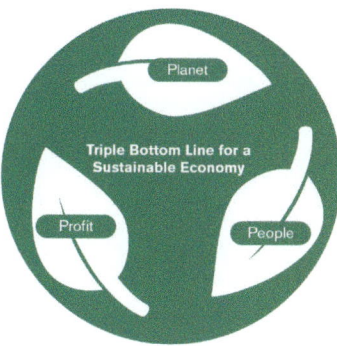

Figure 1.4 The triple bottom line for a sustainable economy includes social (people), environmental (planet), and economic (profit) benefits to developing green buildings.

Social, Environmental, and Economic Benefits of Green Buildings

Social Benefits

- Improved indoor environmental quality
- Healthier environment
- No additional burden on the local infrastructure

Environmental Benefits

- Safeguards wildlife
- Absence of air or water pollution
- No extra waste production
- Prevents the depletion of natural resources

Economic Benefits

- Minimizes operating and maintenance costs
- Higher property value
- Enhances employee productivity
- Lowers life cycle costs (Figure 1.5)

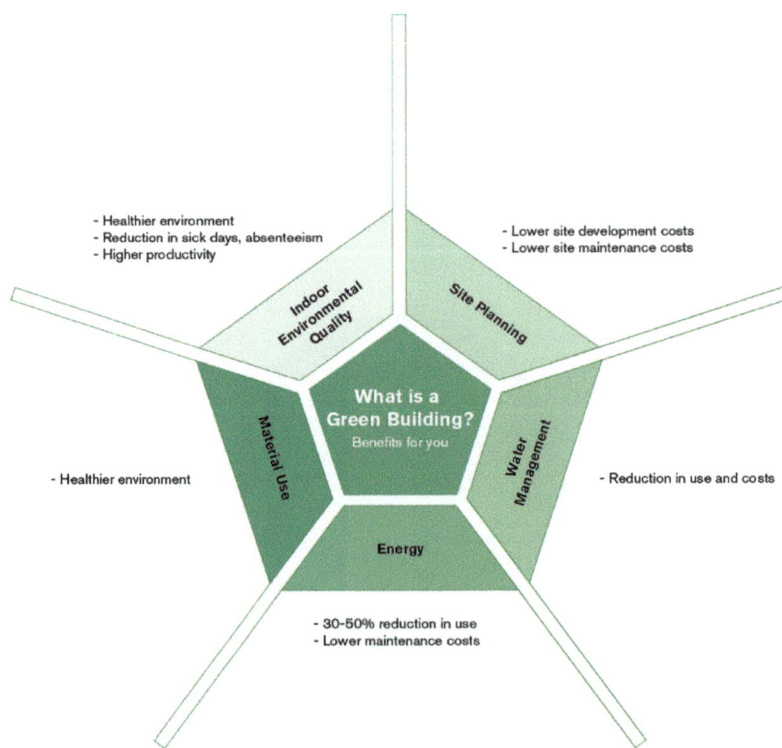

Figure 1.5 Summary of the benefits of green buildings.

The Case for Green Buildings

The most common argument against constructing more sustainably is the additional initial cost attached to building to a LEED standard. While this cost is, in fact, usually estimated between 1% and 5%, that 1% increase on a large 100 million dollar building would come down to a whopping 1 million dollars. Accordingly, one of the keys to the success of the green building movement is facilitating an understanding of the building's life cycle impact. Traditionally, the industry only sheds light on hard costs (construction-based) and soft costs (design based). In the green building industry, we must look at the entire life cycle cost (LCC) of the building over its total lifespan.

As mentioned in the introduction, LEED does not reinvent the wheel but rather utilizes proven concepts and integrates those ideas into its own rating system. One of these crucial concepts is known as the life cycle approach or life cycle assessment (LCA).

LCA is a technique that gauges and analyzes the environmental impact of a service, material, or product throughout its entire life cycle. LCA evaluates a material's life cycle with respect to:

- Energy
- Emissions
- Waste

Life Cycle Cost (LCC)

Assessment of building costs from "cradle to grave" and the impacts are reviewed at each stage of the building's life cycle. Life cycle costs reflect all factors ranging from initial costs, health issues, and productivity to potential building reuse after demolition. Sustainable buildings aspire to turn the idiom "cradle to grave" into "cradle to cradle" as a closed loop as opposed to a destructive open loop, which produces waste (Figure 1.6).

Our Potential Future

We now genuinely understand and are sufficiently aware of the gloomy future and challenges ahead of us if we continue to advance at our current rate of resource consumption and supplemental artificially induced climate change. Hence, it is high time that we take a look at the possibility of pursuing a completely different path altogether. A path in which buildings not only reduce their impact on the planet, its people, and potential profit but rather positively contribute to the triple bottom line.

Buildings may not simply use the required minimum amount of resources but rather be designed and maintained to have an overall net-positive impact on the environment and potentially reverse the damage. Such buildings are also known as regenerative buildings or net-zero/net-positive buildings,

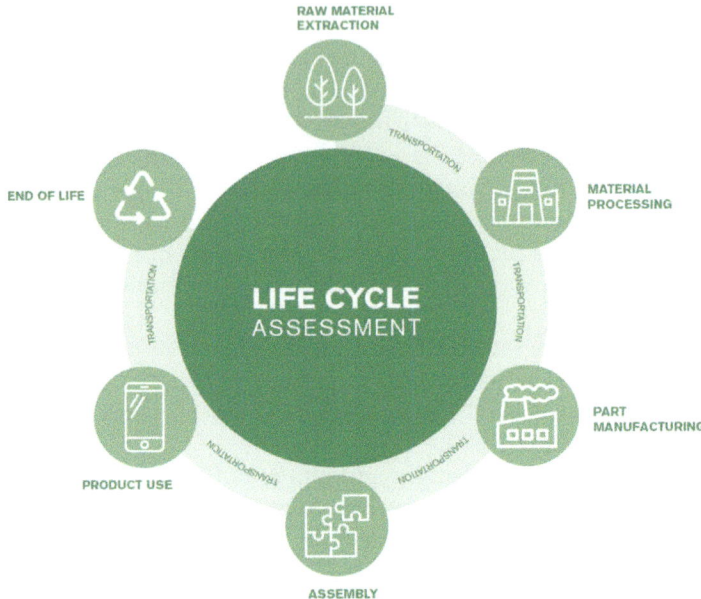

Figure 1.6 Graphic to show the different "cradle to grave" stages involved in life cycle assessment of building costs.

which generate more energy than they consume. Our future can include mixed-use sustainable communities and promote a strong sense of community. The building industry can work with the environment as opposed to battling against nature.

Notes

1 "Learn About Sustainability," EPA United States Environmental Protection Agency, last modified November 14 2022, https://www.epa.gov/sustainability/learn-about-sustainability
2 "What is the greenhouse effect?" NASA, last modified December 16, 2022, https://climate.nasa.gov/faq/19/what-is-the-greenhouse-effect/
3 *Merriam-Webster.com Dictionary*, s.v. "greenwashing," last modified December 7, 2022, https://www.merriam-webster.com/dictionary/greenwashing

2 The Governing Bodies of LEED – USGBC + GBCI

The LEED® Green Associate exam requires a keen understanding of the governing bodies responsible for creating the LEED rating system and operating it effectively to this day. Several students have a tough time answering questions related to this chapter as you cannot rely on logic alone but rather need to depend on remembering the following facts. The LEED Green Associate exam is aimed at testing you on your overall understanding of the LEED rating system. However, to do so, we must first and foremost have a deep and thorough knowledge of its creation and the rationale behind it. This chapter discusses how LEED was created, by whom it was created and how the LEED rating system functions in the industry.

U.S. Green Building Council (USGBC)

The USGBC, or U.S. Green Building Council, is best described as the creator of LEED, which is responsible for updating the LEED rating system and defining what makes a green building truly sustainable. Established in 1993, with its headquarters in Washington D.C., The USGBC continues to exist as a private non-profit organization to this day. They represent companies and organizations from all across the building industry in conjunction with the built environment's design, construction, and financial sides.

DOI: 10.1201/9781003405856-2

USGBC's Mission

"To transform the way buildings and communities are designed, built and operated, enabling an environmentally and socially responsible, healthy and prosperous environment that improves the quality of life."

USGBC's Vision

Buildings and communities will regenerate and sustain the health and vitality of all life within a generation.[1]

The USGBC prides itself on being:

Legitimate – The LEED rating system is conceived by Professionals for Professionals, resulting in a realistic green building rating system that promotes sustainable practices that are within the present industry's reach.

Committed – Often, individual stakeholders, for instance, one designer, who does not share a common culture or belief, may not consider how their decisions would affect the building's short and long-term sustainable impact. However, LEED projects demand all stakeholders think about how their initial and ongoing design and construction decisions relate to the sustainability of the building.

Marketable – Frankly speaking, if the public does not understand the literal meaning or essence of LEED, there will be no or a bare minimum demand for LEED buildings. The USGBC has made a conscious effort to educate the public on the practices required for a LEED building so that they can continue to demand it from developers. If developers aspire to continue to grow, their policies must be aligned with and be adjacent to market demand.

The USGBC also provides:

- **Educational opportunities** for both the public and industry professionals through the medium of online or live seminars.
- **Green building resources**, strategies, and tools for project teams and organizations interested in executing green buildings.
- **Networking** through forums to support green building dialogue and communication.
- Tracks and follows up on the status and progress of all **LEED professionals**, including Green Associates and AP+.
- **Greenbuild** – Since 2002, the USGBC annually hosts the world's largest green building conference, which has expanded rapidly all across the globe.

There are only **two formally acceptable ways** to refer to the USGBC:

1   U.S. Green Building Council
2   USGBC

Green Business Certification Inc. (GBCI)

While the USGBC should be thought of as the creators and architects of LEED, the GBCI is the organization that enforces and meticulously reviews the LEED rating system. The Green Business Certification Inc. (GBCI) is the premier organization that independently recognizes and acknowledges excellence in green business industry performance and practices globally through third-party verification services for certification and credentialing. Founded in January 2008, the GBCI, backed by the USGBC, offers independent oversight and control over professional credentialing and project certification under the Leadership in Energy and Environmental Design (LEED) green building rating system.

Currently, the GBCI is the only certification and credentialing body within the green business and sustainability industry that exclusively administers project certifications and professional credentials of LEED, EDGE, PEER, WELL, SITES, Sustainability Excellence, Parksmart, and TRUE. It is a separate entity that manages and supervises the LEED professional accreditation program and the LEED project certification process by evaluating a project's LEED application. Furthermore, the GBCI also administers the credential maintenance program (CMP), which necessitates all LEED Green Associate and AP+ credential holders to maintain their accreditation in good standing (Figure 2.1).[2]

Leadership in Energy and Environmental Design (LEED)

Leadership in Energy and Environmental Design (LEED) furnishes a framework for healthy, efficient, carbon, and cost-saving green buildings. LEED certification is a globally recognized symbol of sustainability achievement and leadership.

LEED is a holistic system that not only focuses solely and exclusively on a single element of a building, such as energy, water, or health; but rather

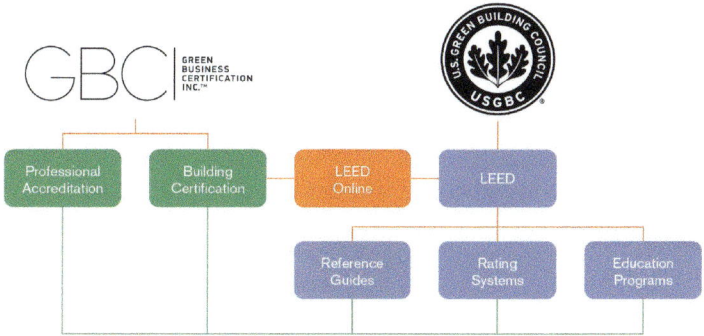

Figure 2.1 The roles of the GBCI and USGBC.

also looks at the big picture, factoring in and acknowledging all the critical elements that work in conjunction to create the best building possible. The LEED rating system is classified into the following:[3]

- 35% of the credits in LEED are related to climate change
- 20% of the credits directly impact human health
- 15% of the credits impact water resources
- 10% of the credits affect biodiversity
- 10% of the credits relate to the green economy
- 5% of the credits influence the community
- 5% of the credits affect natural resources

This will be discussed in detail later in this chapter as a part of the impact categories.

The Success of LEED

The reason why LEED has upheld its position as the dominant market leader in rating green buildings may be broken down into the following categories:[4]

1 **LEED for the developer**:
   a LEED buildings sell and rent at a premium in contrast to similar conventional non-licensed buildings. Tenants are willing to pay more to live or work in healthier, eco-friendly, and green-certified buildings.
   b LEED-certified buildings have higher occupancy rates, highlighted during the 2008 Great Recession, as LEED buildings maintained their occupancy rates better than non-LEED buildings. This conveys that higher quality tenants who can survive economic downturns are prone to select LEED buildings.
   c LEED buildings have lower operational and maintenance costs and are less vulnerable to commodity fluctuations simply by utilizing less water and energy.
   d Depending on the location of the building, there may emerge opportunities for incentives such as lower permit fees, zoning exceptions, and financial incentives.
2 **LEED for the consumer**:
   a It is our responsibility to demand green buildings that are both healthier and beneficial for us as well as the planet.
   b LEED buildings give rise to more comfortable indoor environmental quality in addition to an elevation in employee productivity.
   c We see a generational shift where climate change awareness is spread among the younger generation as a part of their elementary education, and Earth Day and Earth Hour are now fixtures in our calendar.

    d   LEED promotes a brand new job market known as the green-collar industry, lending new employment avenues to passionate individuals who want to be part of the workforce in the emerging environmental sector.

3   **LEED for the environment**:

    a   There is only one earth that provides us with everything we need, and Planet B does not exist. LEED understands the importance of preserving and conserving our home for the coming generations, inhabited by millions of species.

    b   LEED buildings scale down the emissions of greenhouse gases from the building itself, transportation, and manufacturing products we employ throughout the building's lifespan.

Extra brownie points if you noticed the parallels between the above paragraphs and the triple bottom line as LEED attempts to actively address the needs of the environment, society, and economy.

LEED Certification

The term 'LEED certified' is synonymous with the term 'green building.' Being LEED certified indicates that a building has met the green building criteria and achieved sufficient credits in LEED to hold the coveted designation. As mentioned previously, LEED, a universally recognized green building standard, is a point-based rating system where a project must earn a minimum number of points. Buildings can achieve one of the following levels of accreditation based on the number of points they possess. Perhaps you recognize these plaques from your daily life.

A total of 110 points are distributed across the various categories in LEED. However, the minimum required number of points is 40 points. While point-related questions are seldom asked on the LEED Green Associate exam, you must be aware of the corresponding levels of award. A good trick to memorizing and recalling the points assigned to each level is that the '5' looks like an 'S,' which stands for Silver, and the '6' looks like a 'G,' which represents Gold. Nevertheless, do not get confused between the following:

- **LEED-Certified with an uppercase 'C'** refers to a project that has attained 40–49 points and was awarded the basic certification level.
- **LEED-certified with a lowercase 'c'** applies to a building certified to any level ranging from the basic 'Certified' level to the 'Platinum' level, which corresponds to the highest score.

A common industry pet peeve is calling LEED, LEEDS. There is no 'S' in LEED certification. Leeds is a city in the UK and not a green building rating system.

LEED Accreditation

While buildings can become LEED certified, individuals can also strive to become LEED accredited. However, remember that on the one hand, buildings are **certified**; on the flip side, people become **accredited**. There are three levels or grades of individual accreditation that are awarded to those who have the know-how in the green building industry, and they come in the following three professional designations:

- **LEED Green Associate** – The LEED Green Associate Credential is a unique professional designation designed to show both employers and clients alike that you have attained a strong understanding of green building principles and the LEED rating system. Accreditation can be the deciding factor while hiring, as companies desire to have numerous LEED professionals on their teams to enhance their bids on future projects. Passing the LEED Green Associate is a prerequisite to the LEED Accredited Professional.
- **LEED Accredited Professional (LEED AP+)** – The LEED AP+ is more technical in nature when compared to the LEED Green Associate. This exam focuses on the details of the LEED reference guide, the bible of LEED, which encompasses all the rules, points per credit, options, and detailed strategies to achieve LEED credits. While the LEED Green Associate is more generic, the LEED AP+ is intended for professionals working towards a LEED building in any capacity (Figure 2.2).
- **LEED Fellow** – LEED Fellow, is the highest level of LEED accreditation. They are nominated by their peers, need to undergo an extensive portfolio review, and must have a minimum of 10 years of experience in the green building industry while holding a LEED AP with a specialty credential. They have a background and history of exemplary leadership, impactful commitment, service, and advocacy in green building and sustainability.

Credential Maintenance Program (CMP)

The GBCI further developed a continuing education (CE) process known as the Credential Maintenance Program (CMP). In order to renew and maintain a LEED credential, the LEED professional must;

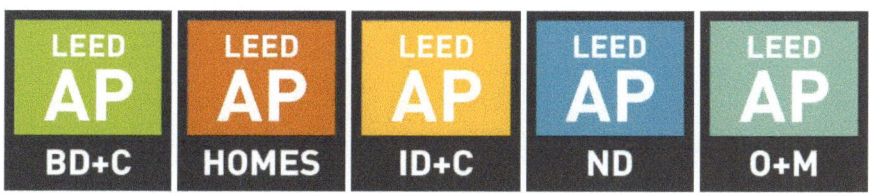

Figure 2.2 LEED variants for type and size of building project.

- Earn CE hours as described in the CMP within a 2-year reporting cycle.
  - LEED Green Associates must earn **15 CE hours** (3 must be LEED-specific).
  - LEED APs with Specialty must earn **30 CE hours** (6 must be LEED-specific), as well as **6 additional hours** for each additional specialty.
- CE hours must be earned and reported during the 2-year reporting cycle after passing.
- All LEED Professionals must pay an $85 renewal fee every 2 years.

LEED Rating Systems

One of the underlying reasons why LEED has thrived in the built environment is its ability to adapt to changes in the market. The USGBC recognized that different buildings required unique attributes because one rating system cannot be compatible with all buildings. As per LEED version 4, the current version of LEED, there are five different LEED rating systems since a new building, and an existing building cannot be judged identically. Within the rating systems, market adaptations exist because a new hospital has completely different requirements compared to a new school (Figure 2.3).

- **LEED for Building Design and Construction (LEED BD+C)** – It focuses on new construction and major renovations of various types of buildings, as illustrated in the diagram below. While each of these market adaptations are similar, they each have a number of credits that are unique to their building type, which we'll explore in the following chapters.
- **LEED for Building Design and Construction – Core and Shell** differ from the other seven market adaptations as it only certifies the building envelope (exterior walls, windows, roof, and floor) and the shared areas such as the lobby and hallways. An example of a Core and Shell building is a developer constructing an office condo tower but does not know who the future tenants/companies will be, who eventually purchase and occupy each space.

The LEED Green Associate exam is primarily based on the LEED for BD+C rating system and, more specifically, the new construction market

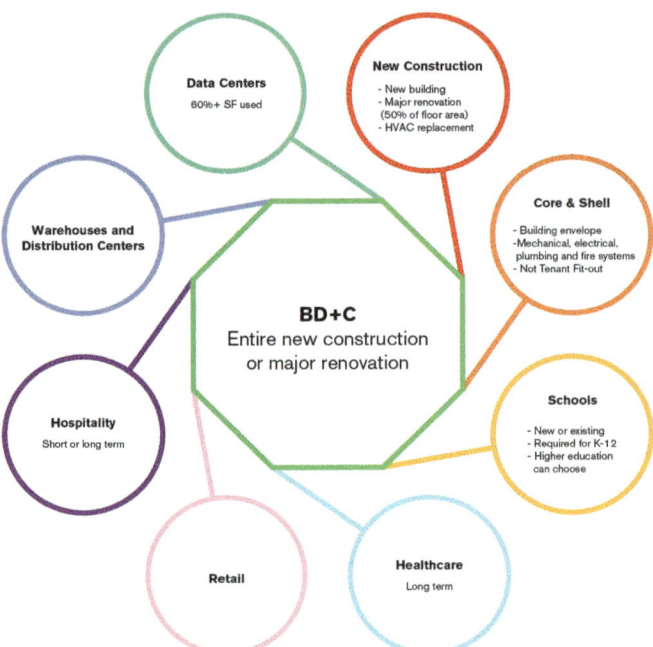

Figure 2.3 Types of building project assessed by LEED for Building Design and Construction (LEED BD+C).

adaptation, which is the most common rating system currently used. However, you must be aware of the existence and function of other rating systems as listed below:

LEED for Interior Design and Construction

LEED for Interior Design and Construction (LEED ID+C) enables project teams, who may not have control over whole building operations, the opportunity to develop indoor spaces that are favorable for both the planet and people (Figure 2.4).[5]

LEED for Operations and Maintenance

LEED for Operations and Maintenance (O+M) offers existing buildings an opportunity to pay close attention to building operations by supporting whole buildings and interior spaces that have been fully operational and occupied for a period of at least 1 year. The project may be undergoing improvement work or little to no construction work. The LEED O+M certification is the only rating system which requires recertification every 5 years after the award is initially achieved (Figure 2.5).

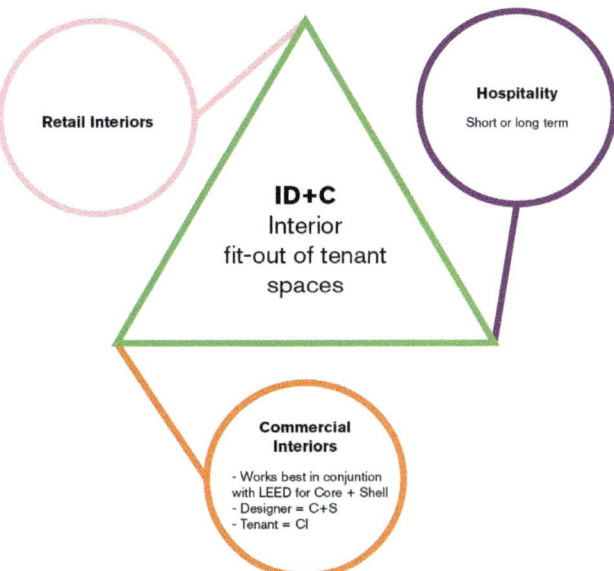

Figure 2.4 Types of building project assessed by LEED for Interior Design and Construction (LEED ID+C).

Figure 2.5 Types of building project assessed by LEED for Operations and Maintenance (LEED O+M).

Multiple Certifications

It is relevant to note that a building can have multiple LEED certifications. For example, a building certified under New Construction in LEED version 1 in 2000 can just as well presently pursue the LEED certification for Existing Buildings Operations and Maintenance (O+M).

Another instance is when a developer certifies the Core and Shell (LEED BD+C) of an office condo tower, followed by an individual tenant who certifies their interior space or office under LEED ID+C.

LEED for Homes

LEED for Homes is geared towards promoting the transformation of the homebuilding industry towards more sustainable practices. It is designed to fulfill the needs of single-family homes, low-rise multi-family (one to three stories), or mid-rise multi-family (four to eight stories) homes.

Compared to the other four rating systems, LEED for Homes is quite the opposite and the most different, while being the only rating system that requires onsite verification. The other four rating systems rely solely on documentation such as calculations, signed contracts, and as-built construction drawings submitted through LEED Online. On the other hand, LEED for Homes counts on third-party verification provided by Green Raters and LEED for Homes Providers.

Image by acongraphic on Freepik

Green Raters provide in-the-field verification services and collaborate with LEED for Homes Provider Organizations to complete the verification process on every single LEED for Homes project. Green Raters must be associated with the project right from the design phase and throughout the construction process. They are responsible for:

- Providing onsite verification of services.
- Assembling the Project Submittal Package and submitting it for certification review.
- Verifying that the home is designed and built to satisfy the requirements of the LEED for Homes rating system.

LEED for Homes Providers provides quality assurance oversight for each Green Rater. LEED for Homes Providers are local organizations or companies carefully screened by the USGBC based on their demonstrated experience managing a team of Green Raters and supporting builders of high-performance homes. They can also play the role of a Home Energy Rating System Rater (HERS Rater). A provider is in charge of the following:

- Recruitment and registration of projects
- Coordination and oversight of Green Raters
- Certification of LEED for homes
- Quality assurance
- Documentation flows from the owner → Green Rater → LEED for Homes Provider → USGBC/GBCI

LEED for Neighborhood Development

LEED for Neighborhood Development (LEED ND) was engineered to inspire and help construct better, more sustainable, and well-connected neighborhoods. It looks beyond the scale of buildings to consider entire communities.

LEED ND strives to push the entire industry towards sustainable development as a community. While focusing only on one building can only accomplish so much on its own, a community that is walkable, mixed-use, and shares its amenities and resources can make a more significant positive impact on the built environment. Check out the Hudson Yards project in New York City as a prime example of LEED ND.

There are two paths that can be pursued:[6]

- **LEED ND Plan**
  - The project is currently in the planning phase, and up to 75% is already constructed.
- **LEED ND Built Project**
  - In the case of projects nearing completion or were developed within the last 3 years.

LEED for ND is comprised of the following credit categories:

1 Smart Location & Linkage
2 Neighborhood Pattern & Design
3 Green Infrastructure & Buildings
4 Innovation & Design Process
5 Regional Priority Credits

LEED Bulk Certifications

There may come a time when a bulk of projects need to be certified simultaneously. For the purpose of preventing a backlog in the certification and reviewing process by the GCBI, LEED offers a bulk submission process for similar projects or projects in the same location, and there are two options when it comes to certifying multiple buildings as jotted below:

- **LEED Campus** is used for multiple buildings on a site under the control of a single entity.
- **LEED Volume** is used for 25 or more identical projects under BD+C or O+M (Ex. Walmart or Target expansion).

Rating System Guidance Selection

In order to determine and conclude which of the five rating systems is appropriate for your project, the 40/60 rule was developed.

THE 40/60 Rule

If a rating system is fit for **more than 60% of the gross floor area (cumulative square footage of each floor)** of a LEED project building or space, then that specific rating system should be employed (Figure 2.6).

In reality, selecting a rating system is usually a pretty obvious and apparent choice. But when numerous rating systems are suitable for a project,

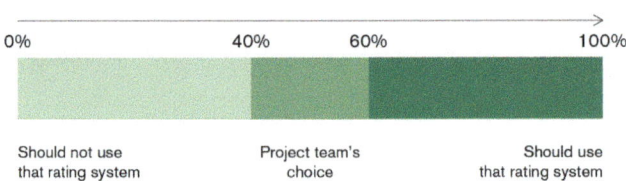

Figure 2.6 Illustration of the 40/60 rule for picking which rating system is appropriate for your project.

Figure 2.7a Example of a mixed construction project.

the 40/60 rule applies. An instance of using the 40/60 rule is by making an addition to an existing building. If the addition consists of more than 60% of the gross floor area of the entire project, then the whole project shall pursue LEED for Building Design and Construction (BD+C).

In the illustration below, it may be observed that the 100,000 sq ft addition makes up over 60% of the total gross floor area of the 120,000 sq ft project, due to which the entire project shall pursue LEED for Building Design and Construction (BD+C) (Figure 2.7a).

In the situation where the 40/60 rule fails to determine which rating system must be selected, ultimately, the decision rests totally on **the project team**.

The entire gross floor area of a LEED project must be certified under a single rating system and is subject to all prerequisites and attempted credits in that rating system, regardless of mixed construction or space usage type.

LEED Rating System Structure

In a LEED rating system, each credit category comprises mandatory **prerequisites and credits** chosen by the project team in conformity with cost, environmental benefit, and organizational goals.

Prerequisites: As the name suggests, prerequisites are the mandatory elements of the LEED rating systems. A LEED project team must adhere to all the prerequisites or set requirements to achieve LEED credentials. If a project fails to comply with any of the prerequisites, it will NOT qualify for certification. No points are awarded for achieving prerequisites.

Credits: Credits are non-mandatory optional elements of LEED rating systems. Project teams in charge of the project choose a combination of credits to score points toward certification. Each credit is worth a certain number of points. If a project meets the requirements of the credit, then it earns the points associated with this credit. Different credits are worth different points, and the point allocation also depends on the degree to which you achieve the credit. The credits may be sorted into two distinct types – credits completed during the design phase and credits achieved during construction.

To earn LEED certification, the project must comply with all the LEED rating system prerequisites and acquire a minimum number of points in the process.

Each LEED prerequisite and credit has an intent, requirements, and strategies to achieve the credit/prerequisite. For the LEED Green Associate exam, the intent is highly crucial, and many questions relate back to the reason behind the credit's existence, while the LEED AP exams focus on a credit's requirements and subsequent point allocation.

The sections mentioned in the LEED rating systems are as follows:

1 **Intents**: This section mentions the sustainability goals and environmental benefits of the credit/prerequisite. Essentially it explains "Why the credit exists."
2 **Behind the Intent**: It describes how the credit fits into the overall sustainability picture.
3 **Requirements**: This section outlines the options or paths to achieve the credit/prerequisite requirements and specifies the number of points associated with the credit.
4 **Step-by-Step Guidance**: They offer general tips and examples on how to implement and document credits.
5 **Further Explanation**: Calculations, special project considerations, and international compliance are displayed in this section.
6 **Related Credits**: Credits/prerequisites may have synergies or tradeoffs between them. This segment lists the other credits/prerequisites affected by achieving this credit/prerequisite.
7 **Referenced Standards**: It furnishes a list of standards, such as ASHRAE, ASTM, and EPA, used as a requirement to achieve the credit/prerequisite. **Federal, state, and local laws and codes** are used if more stringent than these standards.
8 **Changes from LEED 2009**: It depicts and explains in what manner the credit has changed.
9 **Required Documentation**: This section explains and exhibits the necessary documentation to be uploaded to LEED Online and specifies the declarant responsible for signing off on the credit/prerequisite.
10 **Examples**: Examples pertaining to some of the credits/prerequisites on how they were achieved are disclosed.
11 **Exemplary Performance**: Additional points that may be earned from some of the credits for a project that greatly exceeds or doubles performance requirements are outlined in this section. No prerequisite offers

exemplary performance points, and not all the credits have exemplary performance points.

12 **Definitions**: Definitions for terminology specific to this credit/prerequisite can be viewed in this section.

LEED Rating System Credit Categories

Before proceeding any further, it is of utmost importance to introduce you to the categories which make up the LEED rating system, as the Green Associate Exam focuses on their content. All the LEED rating systems, excluding LEED for Neighborhood Development, divide green building strategies into the following categories:

- Integrative Process **(IP)**
- Location and Transportation **(LT)**
- Sustainable Sites **(SS)**
- Water Efficiency **(WE)**
- Energy and Atmosphere **(EA)**
- Materials and Resources **(MR)**
- Indoor Environmental Quality **(EQ)**
- Innovation in Design **(ID)**
- Regional Priority **(RP)**

Figure 2.7b LEED rating system credit categories.

LEED V4 Impact Categories – The Basis of Point Allocation (Credit Weightings)

In consecutive chapters, we will exhaustively scan and investigate inside out the requirements of each credit. You may even wonder why a certain credit is worth many more points than others. The value of each LEED credit is based on how much it satisfies the following impact categories (Figure 2.7b):

1   Reverse Contribution to Global Climate Change
2   Enhance Individual Human Health and Well-Being
3   Protect and Restore Water Resources
4   Protect, Enhance, and Restore Biodiversity and Ecosystem Services
5   Promote Sustainable and Regenerative Material Resources Cycles
6   Build a Greener Economy
7   Enhance Social Equity, Environmental Justice, and Community Quality of Life

For example, the most valued impact category is Reverse Contribution to Global Climate Change, accounting for 35% of credit weighting. Thus, if credit A reverses its contribution to global climate change more than credit B, credit A will likely be worth more points. Note that:

•   All LEED credits are worth a minimum of 1 point.
•   All LEED credits are positive, whole numbers; there are no fractions or negative values.
•   All LEED credits apply to every project regardless of its location (Figure 2.8).

LEED Green Building Reference Guides

A LEED reference guide is an instruction manual that assists project teams through the entire certification process. There is a corresponding reference handbook for each rating system that illustrates each of the credit intents, requirements, implementation strategies, references, etc. LEED Reference Guides equip project teams with all the information they need to achieve certification. It is essentially the LEED code or the bible of LEED.

LEED Rating System Development

The current version of LEED is LEED version 4. All LEED professional exams are based on LEED version 4, which is what this textbook covers. Below, you can catch a glimpse of the steps for creating a new LEED rating system. The first step is improving LEED or making credits more stringent in an effort to push the envelope for more sustainable buildings in the future. This is a key element of LEED, as the industry expects LEED to demonstrate how realistic and achievable green buildings can be. As LEED

Figure 2.8 LEED V4 impact categories.

advances, it can be seen that the local and regional governments adopt prerequisites or credits and include them in their mandated building code. The LEED steering committee consists of industry leaders and academic experts who decide the viability of the proposed changes. Following this review, feedback from all stakeholders and industries regarding the proposed changes is gathered through public comment, and eventually, the new rating system is voted on and approved by USGBC members. There is no set schedule for new versions of LEED, as viability in the market is evaluated annually (Figure 2.9).

Minimum Program Requirements (MPRs)

A LEED project must satisfy Project Minimum Requirements (MPRs). MPRs are mandatory, and they define the minimum characteristics a project must possess in order to be eligible for certification. Even before Prerequisites, MPRs must be met. MPRs are **standard across all rating systems**, and they serve three key objectives:

- Customer guidance and understanding
- Reinforcing LEED's integrity
- Minimizing challenges throughout the LEED certification process

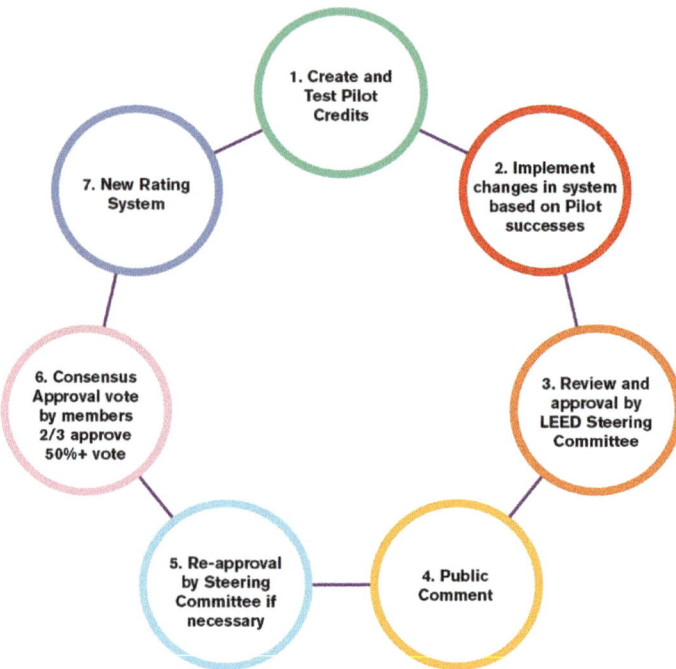

Figure 2.9 Steps for creating a new LEED rating system.

Each rating system has three identical MPRs:

1   Must be in a permanent location on existing land
- Cannot be designed to move.
  - Ex. Trailers or Portables are not eligible
- The land which the building is constructed on must already exist
  - Ex. Building an artificial island in a lake

2   Must use reasonable LEED boundaries
- Includes all contiguous land associated with the project.
  - Ex. Hardscapes (sidewalks and parking), stormwater treatment, and landscaping.
- Must not unreasonably exclude portions of the site which make it easier for the project to meet credits and prerequisites – NO GERRYMANDERING (adjusting the site boundary).
- The Gross Floor Area of a LEED Project **must exceed 2%.**
  - For example, a project site with 100,000 sq ft of land area must contain a building of at least 2,000 sq ft. This can be a single-story building of 2,000 sq ft or a two-storied building of 1,000 sq ft on each floor. This is the bare minimum, and we will explore a theme in LEED of building up as opposed to out and building in developed areas in the LT credit category.

3   Must comply with project size requirements (GFA)
- LEED BD+C and O+M – A minimum of 1,000 sq ft is required.
- LEED ID+C – A minimum of 250 sq ft is necessary.
- LEED ND demands at least two habitable buildings and should be no larger than 1,500 acres.
- LEED for Homes – It is defined as a "dwelling unit" by all applicable codes.
- LEED buildings must be constructed for at least one occupant (Full-Time Equivalent)

Project Registration and Certification

It is often a common sight to see a LEED plaque on buildings in your city. Let's get down to how this award was applied for and achieved. A project pursuing LEED certification must conform to all the Minimum Program Requirements (MPRs), prerequisites, and a minimum number of credits as described in the applicable rating system. Keep in mind that projects must initially always satisfy local, regional, and fire codes, as they are the law! A building that does not comply with the building code cannot be operational.

Registration and Certification

The preliminary step in the accreditation procedure is Project Registration. Projects may be registered on the GBCI website (www.gbci.org), which has information on registration costs which is a flat fee. Registration facilitates access via LEED online website (www.leedonline.com) for communication, software tools, and other critical information. The cost of LEED certification depends on a dollar-per-square-foot basis, where the larger the project, the more there is to review.

LEED Online

LEED online is a web-based tool where the project team takes advantage of managing the LEED certification process. LEED online allows project teams to (Figure 2.10):

- Complete documentation requirements
- Upload required files
- Submit an application for review
- Receive reviewer feedback
- Manage project details
- Earn LEED certification

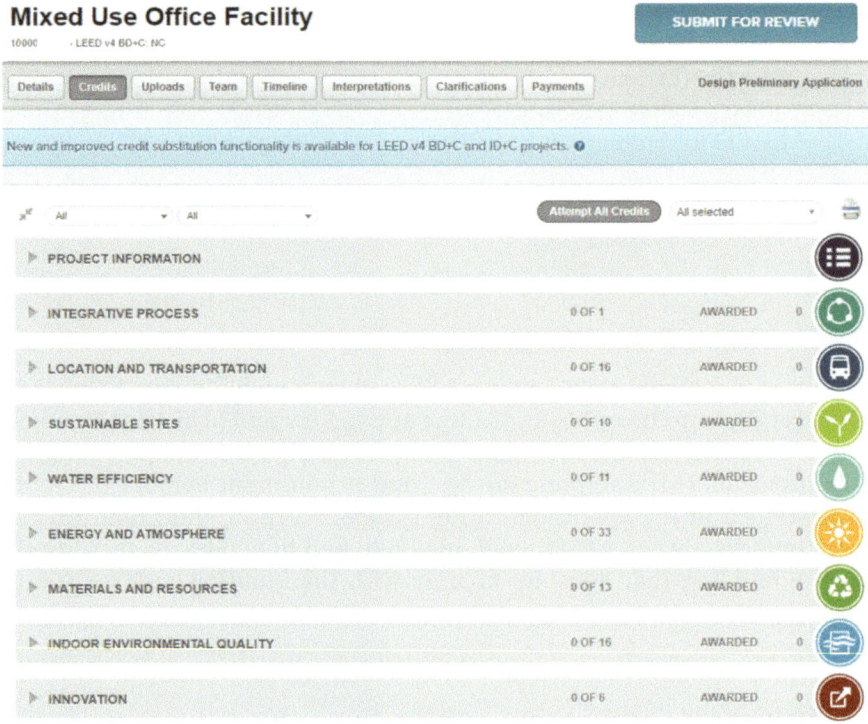

Figure 2.10 Screenshot from LEED online.

Owner and Agent

While the owner controls the property, has authority over its elements, and accepts the certification agreement, the agent, on another note, is granted authority by the owner to register the project.

Project Administrator

The role of the Project Administrator is automatically designated to the person who registers the project via LEED Online by the GBCI. Anyone can easily have a go at registering a LEED project, and the Project Administrator may be allowed to be changed after registration. The Project Administrator's responsibilities include:

- Invite team members to the project on LEED online.
- Assign credits to team members and give them the responsibility to upload/sign credit forms.
- Submit the application for review.
- Accept the reviewer's feedback.

LEED AP

A project with a LEED AP+ who plays a principal role in the project is eligible to receive 1 point under the ID category, provided they have the correct corresponding specialty. A maximum of 1 point can be achieved regardless of the number of LEED APs. It is not at all mandatory to have a LEED AP in the project team in order to achieve certification, but a LEED AP has the following capabilities:

- Understands the certification process.
- Assigns credits to team members depending on their expertise and credit requirements.
- Coordinates between team members from different disciplines.
- Manages the credit documentation and uploads files.
- Coordinates between local codes and standards and LEED requirements.

Eligibility

A project that abides by all the required MPRs and prerequisites, achieving the minimum number of points to earn a certification level, is a candidate for LEED certification.

Project Checklist (LEED Credit Scorecard)

A project checklist is a form provided by the USGBC that covers the prerequisites and credits of the selected rating system. This form is used by the project team (often during a Charrette discussed in Chapter 4) to determine if they can meet all the prerequisites and the certification level they can achieve (Figure 2.11).

The three phases of credit applications are:

1   Predesign – **Discovery**
2   Schematic Design – **Design and Construction**
3   Feedback Mechanisms – **Operations and Performance Monitoring**

Credit Forms and Calculators

Credit forms are utilized to document and verify credit/prerequisite compliance. Credit forms or templates are Adobe interactive PDF forms accessible by Project Administrator and project team members via LEED online. Each credit/prerequisite has its own form, which records the documentation requirements for its achievement and is signed by a specified team member. Calculators are built in for the credits that need calculations.

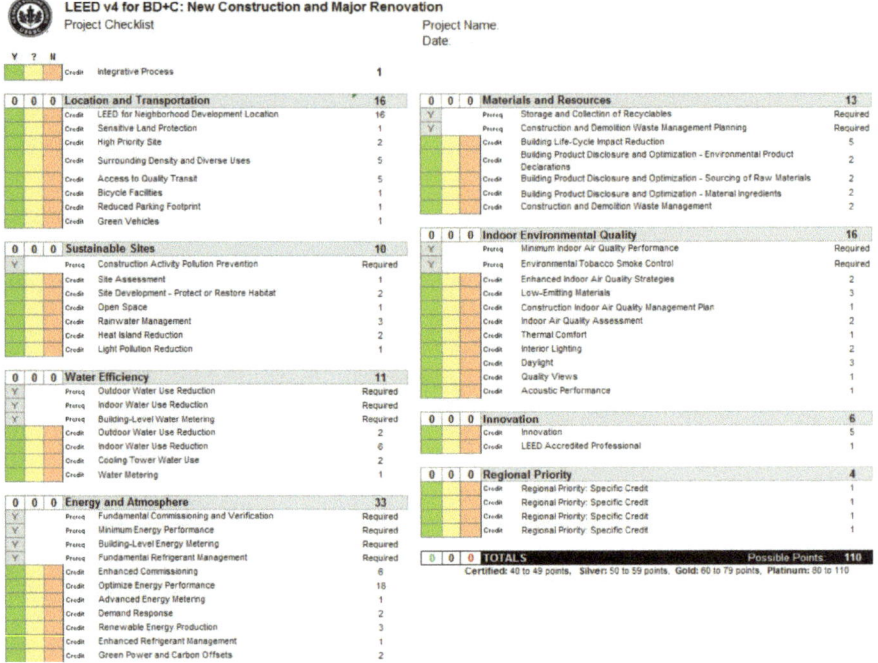

Figure 2.11 LEED credit scorecard.

Samples of Credit Templates and Forms are available on the USGBC website – http://www.usgbc.org/articles/first-look-leed-v4-online-forms

Application Process Outline

1 **Initiate Discovery Phase** – Follow the steps in the Integrative Process Credit.
2 **Select a Rating System** – Use the 40/60 rule to pick the appropriate LEED Rating system.
3 **Check MPR compliance**
4 **Establish Project Goals** – Align with the project's context and the values of the project team, owner, or organization. This is accomplished through a goal-setting workshop (see IP).
5 **Define project scope** – Map the LEED Project boundary and explore any special certification programs such as Volume or Campus applications.
6 **Develop LEED scorecard (see above)** – Select **Y/?/N** for each credit based on expectations.
7 **Continue Discovery Phase** – This phase involves Additional costs and strategy analysis.

8   **Continue Iterative Process** – Repeat all seven steps above until completely satisfied.
9   **Assign Roles and Responsibilities** – One team member leads the process and manages the application and documentation.
10 **Consistent Documentation** – Gather data at regular intervals to ensure ongoing progress.
11 **Quality Assurance** – Review all LEED documentation to avoid errors prior to submission.

Certification Process

In the certification process, there are two options:

- **Split Review**: Some of the project credits/prerequisites may be submitted during the **design phase** as outlined in the LEED reference guide. Conversely, other credits/prerequisites must be presented only during the **construction phase**. However, these points can be earned only after the construction process.
- **Combined Review**: All the credits and prerequisites are submitted for review at the same time. The project team can choose the option that best suits their project case. The combined review is **only available for LEED BD+C and ID+C.**

Review Process

Preliminary Review – Subsequent to this process, the Project Administrator receives a preliminary rating and requests additional information per credit. The Project Administrator has the option to appeal the reviewer's decision.

    Final Review – All new information is reviewed, and the Project Administrator receives a final rating. The Project Administrator has the ability to appeal to the reviewer's decision. However, no further appeals may be made once the decision has been accepted. Once denied, the project cannot reapply for LEED certification.

    Following acceptance of the certification, the project will be incorporated into the LEED Project Directory, and a plaque will be received based on points achieved.

Summarizing USGBC, GBCI, and LEED

- **USGBC** is entirely responsible for developing:
  - LEED rating systems
  - LEED Reference Guides
  - LEED educational courses and research projects

- **GBCI** offers third-party verification for LEED project certification and LEED professional credentials. **GBCI** manages and administers:
  - LEED project certification process from registration to certification
  - LEED professional credentials examination and accreditation
  - The Credential Maintenance Program
- **LEED (not LEEDs)**
  - LEED Accredited Professionals (LEED AP+)
  - LEED Green Associates
  - LEED-certified projects (Certified, Silver, Gold, Platinum)

Notes

1 "Mission and vision," USGBC, accessed November 1, 2022, https://www.usgbc.org/about/mission-vision
2 "About GBCI," GBCI, accessed November 1, 2022, https://www.gbci.org/about
3 "LEED Rating System," USGBC, accessed November 1, 2022, https://www.usgbc.org/leed
4 Eicholtz, Piet, Nils Kok, and John M. Quigley, "The Economics of Green Building," *UC Berkeley: Berkeley Program on Housing and Urban Policy,* September 15, 2010. Retrieved from https://escholarship.org/uc/item/3k16p2rj
5 "LEED Certification for New Interior Spaces," USGBC, accessed November 1, 2022, https://www.usgbc.org/leed/rating-systems/new-interiors
6 "LEED Certification for Neighborhood Development," USGBC, accessed November 1, 2022, https://www.usgbc.org/leed/rating-systems/neighborhood-development

3 Integrative Process (1 Point)

Traditionally, project team members work in isolation. However, quite the opposite, LEED® advocates collaboration and effective interaction through the integrated design approach. This chapter discusses details regarding hosting team meetings at the commencement of the project, known as charrettes, as well as determining the priority sustainable attributes to be included throughout the design and construction of a green building.

In conventional design practices, each project team member operates more or less in a vacuum or void and solely interacts and communicates with the owner. Such an isolated practice can give rise to an inefficient building that misses out on financial, environmental, and social synergies. For instance, if the architect who is in charge of specifying the degree of insulation does not disclose this to the HVAC (heating, ventilation, and air conditioning) engineer, the building may either end up being over-insulated or contain an overly efficient HVAC system. This results in an overdesigned building and subsequently over budget. In LEED, a collaboration or union between these two disciplines ensures that the HVAC and insulation of the building are designed to complement one another, thereby contributing to a cost-effective building that fulfills its sustainable and financial goals.

DOI: 10.1201/9781003405856-3

Integrated Design Approach

This comprehensive approach marks a paradigm shift or radical transformation from the linear planning process to a human-centered iterative approach embodying all project team members, including the property owner, facility manager, designers, and contractors, collectively working together hand in hand as a team in the early stages of the project to set the sustainable goals of the project and promote synergies between the parties.

The Integrated Design Process (IDP) is highly imperative in maximizing sustainable success as all members must cooperate and work together in an open-ended, cohesive team. Whole Building or Holistic Design approaches furnish forward-thinking sustainable designs.

Integrated Project Team

The integrated project team is the critical factor to the success of the IDP. The project team encompasses key stakeholders and professionals from every discipline and requires the following:

- Effective communication and coordination between team members.
- In-depth analysis to develop innovative solutions.
- Collaboration among all members right from the early stages and throughout all project phases.
- Decisions based on information gathered from all members to satisfy the project goals and prevent conflicts (Figure 3.1).

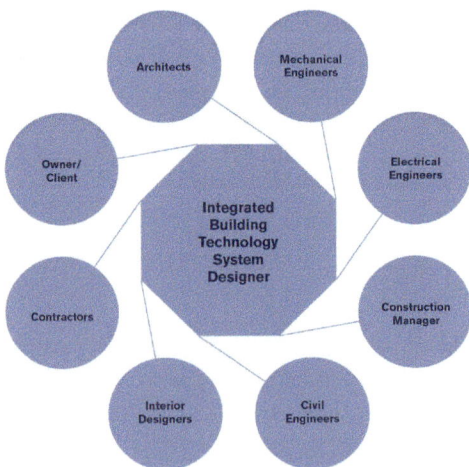

Figure 3.1 Stakeholders involved in the Integrated Project Team.

Integrated Project Delivery (IPD)

Unlike the traditional design/build or design/bid/build project delivery methods, Integrated Project Delivery is employed to promote collaboration among principal stakeholders and design professionals at the **preliminary stages of the design phase**. The owner, architect, general contractor, building engineers, fabricators, and subcontractors join forces and work together throughout the design and construction process. During the initial phase of the process, the project team members **collaboratively define the project's sustainable goals,** and during the course of the project construction process, the technologies and strategies are reviewed to verify adherence to these goals. By keeping a close watch on the graphs, one can observe that the integrative process (IP) results in greater cost control as it is exorbitant to make any alterations late in the project construction phases. Imagine how expensive it is to redesign a half-built building which is more common than you might think. Furthermore, this is also aimed at avoiding green features being value engineered (eliminated to cut costs) out of the project as costs are understood early on by all team members (Figure 3.2).

Design Charrette: A design charrette is an intense, result-oriented brainstorming session held at the beginning of a project to define project goals, swap information, and tackle and confront challenges. In simple terms, it is a pre-design meeting and an indispensable staple of the LEED certification process.

Integrative Process (IP) and LEED

Formerly, IP was simply an actively encouraged route to success in LEED. Nevertheless, with the inception of LEED V4, it is now worth a LEED point. IP is even a prerequisite to be implemented in **healthcare,** and the steps to be undertaken include the following:

TRADITIONAL DESIGN PROCESS

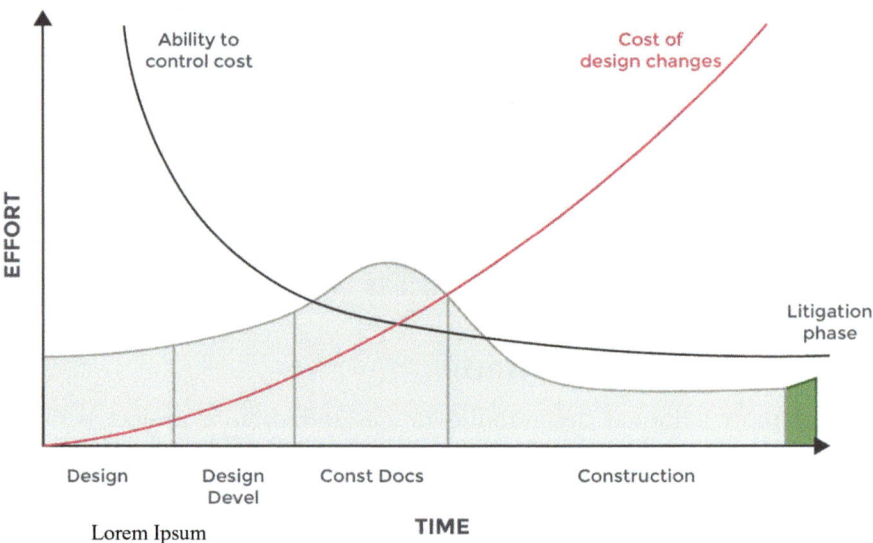

INTEGRATED DESIGN PROCESS

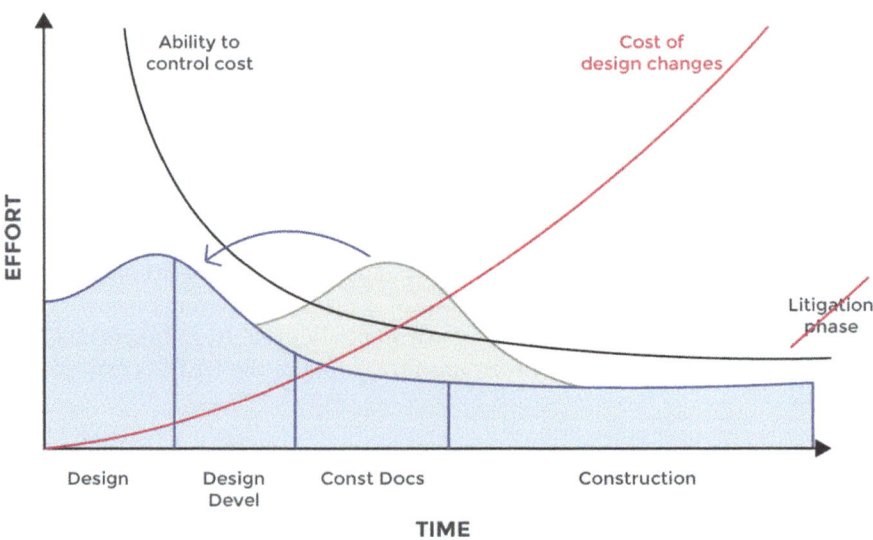

Figure 3.2 Comparison of effort across time involved in the traditional design process and integrated design process.

Identify Team → Prepare and Convene Charrettes → Record Goals

The IP credit applies to all building design and construction umbrella rating systems. The credit prompts and urges the project team to seek and explore opportunities to achieve synergies across all building systems and focuses on energy and water-related systems. The three main phases of IP are (Figure 3.3):

Discovery → Implementation → Performance Feedback

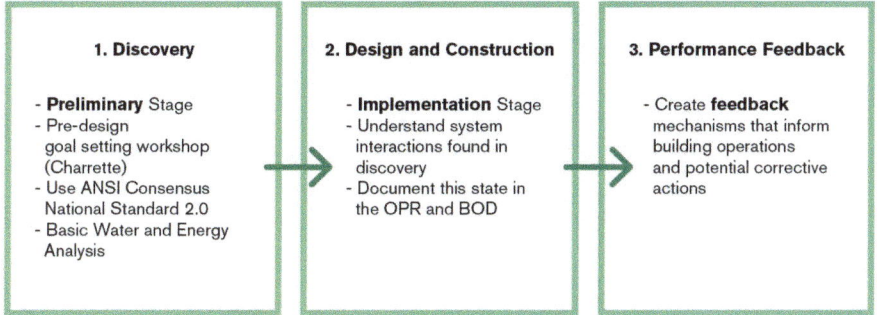

Figure 3.3 Three main phases of the integrative process (IP).

Energy Systems IP Example

1   Discovery – Perform a Simple Box Energy Model Analysis to understand how energy is distributed.
2   Implementation – Make use of the analysis in the discovery phase to create the Owner's Project Requirements (OPR).
3   Feedback – Employ energy meters or a building automation system to ensure proper energy distribution.

Water Systems IP Example

1   Discovery – Conduct a Water Budget Analysis prior to Schematic Design.
2   Implementation – Take advantage of the analysis in the discovery phase to create the Owner's Project Requirements (OPR).
3   Feedback – Utilize water meters to ascertain there are no leaks and there is proper adherence to the OPR.

Objectives of the holistic or integrative design process include credit interactions, high Cost-Benefit Ratios, accessibility, security, synergies, productivity, functionality, and an aesthetically pleasing design.

LEED Points

In a LEED rating system, each credit category is awarded a certain number of points. The IP credit category is unique owing to the fact that it only contains **one credit** worth **one point**. Note that the IP is a prerequisite in LEED version 4 for healthcare (IE Hospitals).

4   Location and Transportation (LT)

Overview

Way Things Were Done in the Past/Problems

It has been forecasted that in the year 2050, 68% of the global demographic will reside in urban areas. At the present moment, transportation emissions are ranked the second largest contributors of greenhouse gases, as 27% of our greenhouse gas emissions are on the grounds of traveling from point A to point B.[1] When coupled with 38% of our greenhouse gas emissions resulting from building operations, it adds up to nearly 70% of global emissions.

How LEED® Addresses It

LEED's location-based solution compels developers to sensibly decide upon their site in a manner that does not damage the surrounding environment, rehabilitate and revive challenging sites, and enhance density to restrict urbanization.

DOI: 10.1201/9781003405856-4

Emissions from transportation systems may be alleviated by adopting diverse approaches, thereby persuading people to give up driving alone in a gasoline-fueled automobile for our future building. The alternative modes include fostering the use of public transit by encouraging and promoting carpooling, bike riding, and facilitating the use of electric vehicles. (16 points).

What LEED Requires for the Credits

LEED requires comprehensive information about the type of site to ensure it is neither sensitive nor contaminated. The surrounding density and proximity to walkable essential services are also gauged.

LEED strongly requests proof of local public transit route frequencies, bike parking, and showers, as well as evidence encouraging carpooling, minimizing parking, and green vehicles.

Location and Transportation (LT) is a unique category as it contains no prerequisites, which basically implies that it may be skipped if required. That being said, the LT category can achieve up to a maximum of 16 points based on the site's location and exactly how occupants will travel to it. Sometimes, a project team is given a site they are stuck with, but LEED does not want to disqualify this common scenario.

Before delving deep into this topic and discussing how to earn points in this credit category, we must be aware of a few terms:

LEED Project Boundary is the portion of the project site submitted for LEED certification, and it must remain consistent for all required credit calculations. If a piece of land supports day-to-day building operations but is offsite, that piece of land must also be included in the LEED boundary. For multiple building developments, the LEED project boundary may be a reasonable portion of the development determined by the project team. For example, in a campus project or an industrial complex, 100% of gross land on campus must be included in LEED boundaries, provided all the campus buildings are LEED certified.

Gerrymandering of a LEED project boundary is prohibited. In short, the boundary may not unreasonably exclude sections of land to create boundaries for the sole purpose of complying with credits.

Building footprint is the area defined as the perimeter of the building plan, i.e., it is the area on the project site used by the building construction. Non-building facilities such as pavements and landscaping are not covered in the building footprint.

Development footprint is the total area on the site where development has taken place, in addition to the pavements, parking lots, landscaping, roads, and other facilities, as well as the building itself. It essentially includes all alterations and modifications made to the site.

Property boundary is referred to as the total area within the legal boundaries of the site (Figure 4.1).

Figure 4.1 Example of a LEED project boundary for a multiple building development.

Location

While selecting a site, the natural and social characteristics of the site should be taken into account, in addition to the existing infrastructure. One must examine how the site is connected to its surroundings and what services are available to the building users. Site selection is paramount as a good site can influence the energy conservation strategies adopted, landscaping and vegetation, and access to public transportation. All LEED rating systems address principles of Smart Growth.

Smart Growth is an urban planning theory that encourages:

- The protection of underdeveloped land
- Transit-oriented and bicycle-friendly land use to minimize the use of automobiles

- Mixed-use development
- Advancement of compact urban centers
- Design of walkable neighborhoods

LEED Credit – Neighborhood Development Location (LTc1)

It is possible for a project to earn all the points available in the entire LT credit category by selecting a site located within the LEED project boundary of an existing LEED for Neighborhood Development certified property. For instance, if a building were constructed on an empty plot of land within the Hudson Yards LEED Neighborhood of Manhattan, this new building would procure 12 points since Hudson Yards was certified as a gold LEED ND project in 2019. Different points are awarded depending on the level of certification. However, no other LT credits can be pursued if this credit is achieved. By building within a LEED ND location, reduced automobile dependence, walkability, existing green infrastructure, and overall health enhancement are guaranteed, and as a result, the next seven credits cannot be obtained. Note that this credit will be more accessible in the future as more LEED ND projects are developed.

Points increase depending on the level of ND certification.

- LEED ND Certified – 8 points
- LEED ND Silver – 10 points
- LEED ND Gold – 12 points
- LEED ND Platinum – 16 points

Selecting a Site

Site selection is the preliminary step in development and can help obtain many LEED points. This is a compelling argument for why LEED must be factored into and acknowledged in the initial stages of development. LEED favors and promotes the growth of previously developed sites for the purpose of safe-guarding the natural habitat, trees, streams, and native plants or species.

LEED Credit – Sensitive Land Protection – LTc2 – 1 Point

The basic rationale behind this credit is to preserve and shield sensitive lands and steer clear of destroying and ruining untouched areas by reusing sites that were developed in the past.

There are two options when it comes to satisfying this credit.

A project team can either pinpoint the development conducted on a **previously developed site – or –**

A project team seeking LEED certification must refrain from developing on the below-mentioned land types:

1   **Prime farmland** defined by the Natural Resources Conservation Service (NRCS) Soil Survey
2   Lands situated within **100 ft of water bodies** as defined by the US Army Corps of Engineers Wetlands
3   Lands located within **500 ft of wetlands** as regulated by the US Army Corps of Engineers Wetlands
4   Areas that are a **habitat for threatened or endangered species** as defined by US Endangered Species Act or NatureServe
5   **Floodplains** as defined by the Federal Emergency Management Agency (FEMA)

Keep in mind that this LEED credit is worth 1 point. Despite the possibility that a building has been constructed on prime farmland, it can still attain LEED certification, as there are 109 other points to achieve the 40-point minimum.

LEED Credit – High Priority Site – LTc3 – 1–2 Points

At times LEED credits offer points to incentivize project teams to undertake a challenge such as a high-priority site. Typically, the greater the challenge, the higher the reward via LEED points. The three terms we must be mindful of to digest the credit's intent and requirements are:

- **Brownfield** – A brownfield site is a previously used or developed land that is presently unoccupied and may have a high degree possibility of being contaminated with hazardous waste or pollution and has the potential to be reused once any toxic substances, pollutants, or contaminants are remediated, and the land is decontaminated.
- **Historic District** – Historic District refers to a group of buildings/structures considered historically significant to the area. It generally includes additional rules and regulations for the proposed buildings with the goal of maintaining the area's character.
- **Infill site** – An infill site is a previously developed site that has been graded, or is a site located between existing building structures. It is essentially a gap in the built environment. Building on infill sites protects undeveloped land and can benefit from the existing infrastructure, such as roads, utilities, and other services. Seventy-five percent of the site must have been previously developed.

This type of credit comprises three options to promote the redevelopment of sites/areas deemed undesirable through decontamination, gentrification, or preservation, and they are:

Option 1 – Historic District – Locate the project within a historic district or infill site.

Option 2 – Priority Designation – Locate project within a site identified by a government agency as for priority redevelopment.

Option 3 – Brownfield Remediation – Select a site that contains soil or water contamination deemed by the federal, regional, or local authority. Though land remediation is both costly as well as time-consuming, a project can also take advantage of existing amenities and infrastructure and potential financial incentives.

LEED Credit – Surrounding Density and Diverse Uses – LTc4 – 1–5 Points

This credit aims to conserve land and habitats by developing in areas where existing infrastructure exists. Favoring pedestrian walkability drastically reduces automobile trips by developing in a dense area with a lot of diversity. This credit persistently continues a theme in LEED of building up as opposed to out and developing where buildings and communities are currently in place. LEED goes to great lengths to discourage the use of privately driven gasoline-fueled automobiles and foster sharing of infrastructure and amenities.

There are two options to demonstrate compliance with this credit that sheds light on density and diverse uses.

Buildable Land – Land where construction can occur and excludes the public right of way (i.e., Public roads and sidewalks)
Density – A measure of floor area (square feet) per unit per buildable land (acre)
Floor Area Ratio (FAR) – Density of non-residential land-use

$$FAR = \frac{\text{Building Floor Area (SF)}}{\text{Buildable Land Area}}$$

Option 1 – Surrounding Density

In this option, LEED looks at the density of the surrounding quarter-mile radius to ascertain the number of points rewarded. The more buildings in the surrounding area, the more people occupy them, and these occupants are serviced by infrastructure and amenities shared or contributed by the proposed new building. Typically, the greater the density, the greater the point reward (Figure 4.2).

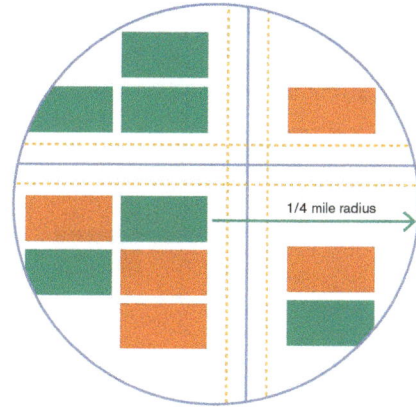

Figure 4.2 LEED gives credit to building projects located in areas of high property density.

Option 2 – Diverse Uses

This option may be considered as a building's walk score. Here, LEED ignores the surrounding area's density but pays attention to the types of services that already exist. If a building must be located in a financial district surrounded by ten retail banks, it does not give its occupants much choice when it comes to basic services. Instead, it would be far more preferable if there was one bank and a variety of other common amenities.

This option rewards a site that is within ½ mile of walking distance (entrance to entrance) of the following existing and readily accessible diverse uses such as: (Figure 4.3)

1   **Category**: Food Retail
    Type – Supermarket, Restaurant
2   **Category**: Services
    Type – Bank, Theatre
3   **Category**: Community Retail
    Type – Convenience Store, Pharmacy
4   **Category**: Civic & Community Uses
    Type – Senior Center or Child-care
5   **Category**: Education Facility
    Type – University

A project may only take into account two types per category but must represent three categories for a minimum of four diverse space uses.

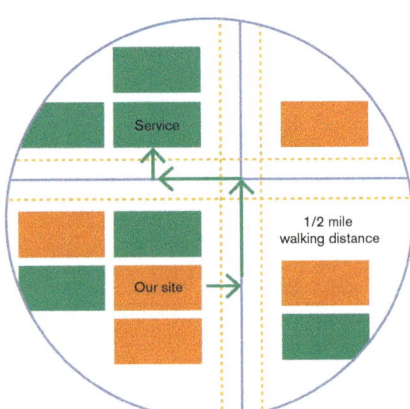

Figure 4.3 Projects receive credit when within a 1/2 mile walking distance from an existing service.

Sample Excerpt from a Project

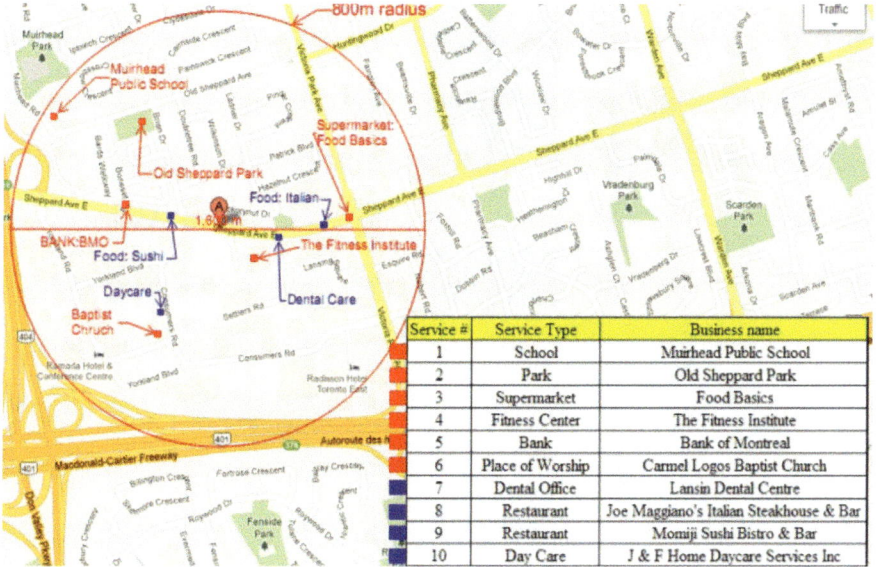

Service #	Service Type	Business name
1	School	Muirhead Public School
2	Park	Old Sheppard Park
3	Supermarket	Food Basics
4	Fitness Center	The Fitness Institute
5	Bank	Bank of Montreal
6	Place of Worship	Carmel Logos Baptist Church
7	Dental Office	Lansin Dental Centre
8	Restaurant	Joe Maggiano's Italian Steakhouse & Bar
9	Restaurant	Momiji Sushi Bistro & Bar
10	Day Care	J & F Home Daycare Services Inc

Figure 4.4 Sample excerpt from a project.

Transportation

A multitude of factors induces the demand for transportation. Additional buildings equal higher transportation demand (Figure 4.4). Studies indicate that transportation accounts for 27% of our greenhouse gas emissions owing to traveling from building A to building B.[2] Under the circumstances, the biggest predicament would be if everyone started driving solo in gas-powered large and heavy vehicles. LEED attempts to combat and resolve this scenario by promoting the development of sites in the vicinity of mass transit and encouraging the use of more efficient modes of transportation. Transportation-related credits in LEED strive to reduce vehicle trips and advocate sustainable alternatives.

LEED Credit – Access to Quality Transit – LTc5 – 1–5 Points

This credit aims to encourage development in locations that possess multimodal transportation choices or reduced motor vehicle use as an alternative. Come to think of it, if a proposed office building is located next to a streetcar or subway stop, the employees will, in all probability, try to cash in on public transit daily to commute to and from work. Doing this would result in those employees personally residing in proximity to public transit stops and not owning a car, as is evident in big metropolitan cities. Nonetheless, this credit mandates a minimum trip count since if only one bus route passes a bus stop twice a day, the occupants would not be left with many options and, as a result, would resort to going the route of owning an automobile.

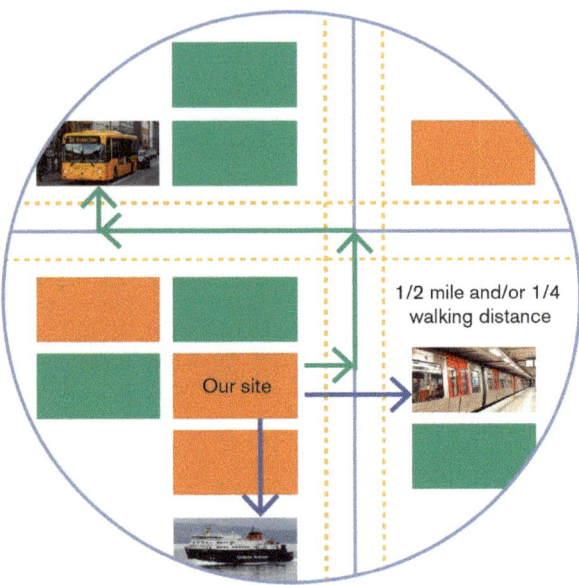

Figure 4.5 LEED gives credit to building projects within 1/4 mile walking distance
        of a bus stop, streetcar stop or rideshare location, or 1/2 mile walking
        distance of a bus rapid transit stop, rail station or commuter rail or ferry
        terminal.

In order to meet this credit's requirements, there are two options (Figure 4.5):

**Option 1 – Ensure that any functional entry is within ¼ mile walking distance
    of either:**
    a   A Bus Stop
    b   Streetcar Stop or
    c   Rideshare Location
Minimum trip count: 72 weekday and 40 Weekend trips
**Option 2 – Ensure that any functional entry is within ½ mile walking distance
    of either:**
    d   Bus Rapid Transit Stop
    e   Light or Heavy Rail Station or
    f   Commuter Rail or Ferry Terminals
Minimum trip count: 24 weekday and 6 Weekend trips
**Demonstrated below is a sample project in which building (C) is located within
    a ¼ mile (400 m) of a bus stop (B or D) serviced by routes in both direc-
    tions and conforming to the minimum trip count. One of the plus points of
    LEED documentation, or how credit compliance is proved, is the practical-
    ity of the evidence. The screenshot captured from google maps below with
    the correct scale showcases the summary and the trip count collected from
    the city's public transportation website (Figure 4.6).**

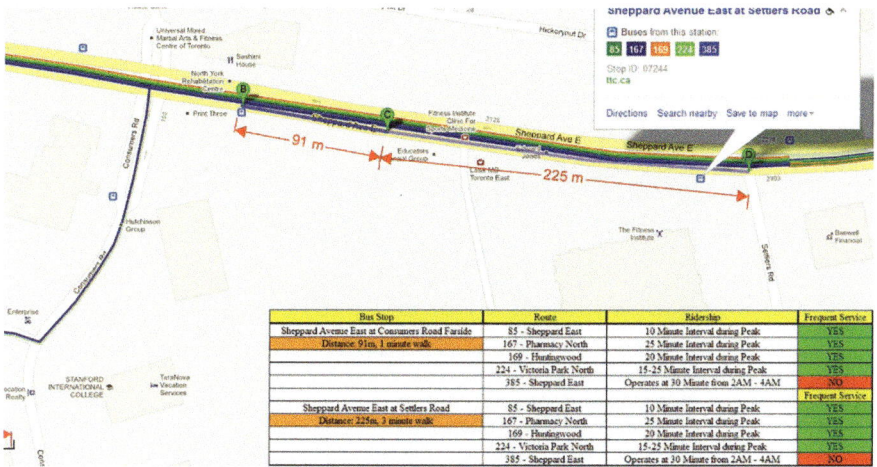

Figure 4.6 Demonstrated above is a sample project.

School Specific Option

Previously, we looked at different market adaptations within the LEED for Building Design and Construction rating system. An example of an option exclusively made available to Schools is ensuring that a certain percentage of students live within ¾ mile walking distance (Grade 8 and below) and/or 1.5 miles (Grade 9 and above) of a functional entry.

LEED Credit – Bicycle Facilities – LTc6 – 1 Point

Another eco-friendly alternative to the car is the bicycle. Cycling significantly reduces CO_2 levels, lowers the risk of disease, and minimizes road congestion to a great extent. This credit demands that adequate bicycle storage solutions be provided for short-term and long-term visitors, as well as a well-connected bike network to alleviate and enhance the conditions for bike riders. However, bicycle parking alone would not suffice if one were to bike to work on a hot summer day. Shower facilities must also be extended to ensure a level of comfort. This credit requires the following:

1   The **bicycle network** must be within **3 miles** of one of the following:
   a   **Ten diverse uses**
   b   A **school or employment** center (if residential project)
   c   **Bus Rapid Transit**, **Light or Heavy Rail Stations**, **Commuter rail, or ferry terminals**
2   Provide short and/or long-term **parking for occupants**
   a   **Long-term** = Covered parking that is easily accessible to residents and employees
   b   **Short-term** = Uncovered Parking used by visitors
3   Provide **Showers for occupants** in commercial spaces such as offices.

The three alternatives mentioned above must be met in this credit to be awarded the point. A helpful frame of mind assumes everyone in the world is lazy, but if the project is designed and located in a certain way to make biking or riding public transit the easiest option, people are sure to change their behavior.

LEED Credit – Reduced Parking Footprint – LTc7 – 1 Point

The US has twice as many parking spots as people! Most of these spaces are impervious (water cannot infiltrate) and above ground. This credit intends to lower automobile dependency and impervious pavements, which is severely detrimental and highly harmful to our environment due to rainwater overflow and contribution to the heat island effect (to be discussed in Sustainable Sites). This credit has three requirements:

Wikimedia Commons user: Xnatedawgx, "Carpool Vehicles Only parking sign at South Transit Center, Fort Collins, Colorado" / CC BY-SA 4.

1 **Do not exceed local zoning code requirements** – Essentially, do not build more parking spaces than required by the law. Plenty of parking spots encourage the use of single-driven automobiles.
2 **Reduce Parking capacity from Baseline (ITE Handbook Standard)** – This is another metric to restrict single-driven automobiles further.
3 **Provide 5% preferred carpool parking**
   a Preferred parking is parking closest to the entrance, excluding accessibility parking spots.
   b A signed agreement by the parking management must be initialed and enforced throughout the life of the building.

LEED Credit – Green Vehicles – LTc8 – 1 Point

When it comes to automobiles, the options are endless. On top of gasoline-powered automobiles, car owners have a multitude of options ranging from hybrids, electric vehicles, and alternatively fueled vehicles that are rapidly gaining viability and, in turn, popularity. LEED refers to the American Council for an Energy-Efficient Economy (ACEEE) to define a green vehicle.

The two requirements for this credit are:

1 **Provide 5% preferred parking or 20%+ discount for green vehicles**
   a American Council for an Energy-Efficient Economy (ACEEE) score of 45+
   - AND EITHER -
2 Provide **Electric Vehicle Charging Stations**
   - OR -
   Provide **alternative refueling/battery switching stations**
   Schools Only – All buses must meet prescribed emission standards, and all other school-owned vehicles must be green.

Noteworthy Standards

• Natural Resources Conservation Service (NRCS) Soil Survey
• US Army Corps of Engineers Wetlands

- US Endangered Species Act
- NatureServe
- Federal Emergency Management Agency (FEMA)
- American Council for an Energy-Efficient Economy (ACEEE)

Notes

1 "68% of the world population projected to live in urban areas by 2050, says UN," *United Nations,* May 16, 2018, Accessed November 1, 2022, https://www.un.org/development/desa/en/news/population/2018-revision-of-world-urbanization-prospects.html
2 "Fast Facts on Transportation Greenhouse Gas Emissions," EPA United States Environmental Protection Agency, Last modified July 14, 2022, https://www.epa.gov/greenvehicles/fast-facts-transportation-greenhouse-gas-emissions

5   Sustainable Sites (SS)

Overview

How Things Were Formerly Done/Problems

Back in the day, little did developers think about the harm they committed to the earth's ecosystems; in fact, they completely ignored it. Deforestation, soil erosion, drop in water table levels, extinction of species, and rivers that run dry and no longer lead to the sea due to overuse are all serious concerns that can be partially attributed to the built urban environment.

How LEED® Addresses It

LEED's sustainability sites category urges developers to reduce pollution on their selected sites by tackling and controlling soil erosion, waterway sedimentation, and airborne debris. This chapter explores six different approaches to enable developers to cut down on pollution. This includes carrying out an adequate and thorough site assessment, safeguarding, and working around natural habitats to promote biodiversity, and maximizing open space, which supports interaction with the environment, reducing run-off, and minimizing the effects of heat islands. (10 points).

DOI: 10.1201/9781003405856-5

What LEED Requires for the Credits

For the sake of attaining credits, LEED requires a detailed site assessment and proof of how the site maximizes vegetation and accessible outdoor space. Likewise, LEED demands proof of increasing rainwater infiltration, reducing the heat island effect by maximizing bright surfaces and vegetated roofs, as well as lowering uplight and glare to minimize light pollution on site. A project's site is considered sustainable in the event it scales down construction impacts, rehabilitates degraded or contaminated areas, preserves water quality, and mitigates rainwater runoff and light pollution. This chapter extensively addresses in depth the Sustainable Sites (SS) category and the strategies selected to achieve it.

SS interact with the Location and Transportation credit category since SS also pivots on the chosen location, though it focuses on the existing site conditions rather than its surroundings. Developers are encouraged to design and construct with the aid of the site's natural elements, contrary to upsetting and disrupting it and fighting nature. A sustainable site design foresees and envisages the entire development footprint, including landscaping and hardscapes. The objective of the site design is to assess the site's ability to support the building while minimizing its ecological repercussions and implications.

LEED Prerequisite – Construction Activity Pollution Prevention – SSp1 – Mandatory

All projects are mandatorily required to comply with all prerequisites to be LEED certified, of which Construction Activity Pollution Prevention is the first prerequisite that must be accomplished at any cost. Bear in mind that if the prerequisite is not met, the project can neither pursue sustainable site credits nor can it earn LEED certification whatsoever.

It is immensely significant to alleviate environmental impacts during the construction process. Construction activities result in:

- The loss of the topsoil
- Loss of nutrients, soil compaction, and decreased biodiversity.
- Water contamination due to water runoff which carries pollutants and sediments to receiving water.

- Airborne dust, which gives rise to environmental challenges and health issues.
- The degradation of water bodies and aquatic habitats.

The project team must devise and implement an Erosion and Sedimentation Control Plan (ESC) to help with construction pollution abatement. The ESC Plan must comply with the **2012 EPA General Construction Permit or local codes,** whichever is more stringent. This is an example of LEED piggybacking existing and proven standards and not reinventing the wheel. LEED finds existing best practices created by experts and incorporates them into the LEED rating system.

Strategies to Control Erosion and Sedimentation

- Temporary or Permanent **Seeding** to stabilize the soil
- **Mulching** using hay, grass, or gravel to hold the soil
- **Earth dike** to divert runoff into sediment traps
- **Straw bales**
- **Silt fence**
- **Erosion control blankets**

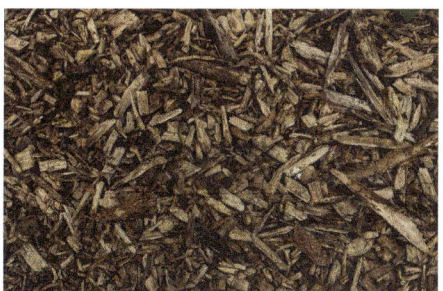

It should be noted that Schools and Healthcare projects are required to conduct a Phase 1 Environmental Site Assessment.

Subsequent to Phase 1, if there ever occurs a situation where contamination is suspected, a Phase 2 Environmental Site Assessment must be performed (SSp2) for the sake of hospital patients and children who are more vulnerable and susceptible to the adverse and negative repercussions of a contaminated site.

LEED Credit – Site Assessment – SSc1 – 1 Point

Site Assessment is the first credit which is an integral part of the integrative process. Rationally, this credit is engaged in determining and defining existing site conditions. The underlying notion behind this credit is that for a project team to derive benefit from and take full advantage of the site, they must make it a point to profoundly understand its contents and learn all that's there to learn about the site. This is one of the most straightforward credits that helps a project team foresee, identify, and plan for challenges closely associated with the site itself. This credit insists that the natural conditions should be assessed and include the following:

1   **Topography** – Contour mapping and slope stability risks.
2   **Hydrology** – Flood-prone areas, delineated wetlands and marshes, and other bodies of water.
3   **Climate** – Sun exposure and radiation, heat island effect potential, sun angles, winds, precipitation, and temperature.
4   **Vegetation** – Plant varieties, tree mapping, threatened or endangered species, and unique habitat.
5   **Soils** – Prime farmland, healthy soil, previous development, disturbed and unstable soil.
6   **Human use** – Viewpoints, transportation infrastructure, adjacent properties, and material effects.
7   **Human health effects** – Proximity to vulnerable populations and vicinity to sources of air pollution.

LEED Credit – Site Development: Protect and Restore Habitat –
SSc2 – 1–2 Points

The modern human habitat can be construed as the most destructive, threatening, and toxic lifestyle on the planet. It is our moral responsibility to minimize and revive the natural habitat of a project site while designing and constructing the built environment. Biodiversity, in addition to native and indigenous species, must be held fully accountable for their symbiotic and intrinsic relationship with nature, as our lifestyle totally depends on nature thriving. This credit is committed to preserving the prevailing natural features and endeavors to restore them to their former state and retrieve what was originally on the site.

This credit offers two options, as mentioned below:

1   **1A. Protect** 40% of greenfield area (if greenfield exists)
    For instance, when building on a site that humans have not previously developed, do not make any developments on 40% of the site's square footage. Building a four-story structure contrary to a two-story structure of the same gross floor area is an optimal strategy while continuing on LEED's theme of building up in place of building out.
    - AND -
    1B. Restore 30% of developed land (including building footprint) with native/adaptive vegetation
    As an illustration, when building on an existing lot line to lot line parking lot, the site restoration should be completed on at least 30% of the site area using native or adaptive vegetation which are indigenous species to the land and can naturally thrive.
    - OR -
2   Provide financial support ($0.40/sq ft) for the total site area to a **Land Trust Alliance organization within the same EPA eco-region**
    • **Land Trust** – A non-profit organization actively engaged in conserving and protecting land through conservation easement or acquisition.
    • It is possible to **purchase a few credits** in LEED directly, and this is one among them. However, it does not come at a cheap price, as a 100,000 sq ft site would require a staggering $40,000 financial commitment to a land trust.

LEED Credit – Open Space – SSc3 – 1 Point

Open space provides a habitat for vegetation and wildlife. While on the one hand, it lowers the urban heat island effect and escalates stormwater infiltration, on the flip side, it connects humans to the outdoors, which enhances and strengthens our daily life in terms of productivity and overall well-being. Open Space is equal to the property area, subtracted by the development hardscape footprint, and is basically the vegetated land on site, which is both useable and accessible. A compact high-rise building with the same square footage as a low-rise building significantly reduces the building footprint and minimizes the disturbance of the existing ecosystem. Contingent on open space set aside, this credit requires that:

30% of the site area (including the building footprint) must be open space.
-   **25%** of that open space must be vegetated or have a **tree canopy.**
-   Outdoor space must be **accessible and useable** for building occupants.

These spaces can be employed for a variety of purposes, such as social and physical activities and community gardens. Open space can be as simple and uncomplicated as a walking path leading to the courtyard or benches positioned around a pond, which is done to encourage human interaction with the outdoors.

A Note on Vegetative Roofs for SSc2 and SSc3

Projects that accomplish a density of 1.5+ floor-area ratio (FAR) may include **vegetated roofs** that can be used towards the minimum 25% vegetation requirement. Essentially, LEED does not want to reward a short one-story building with a green roof when it could be two-storied and take up half the land.

LEED Credit – Rainwater Management – SSc4 – 2–3 Points

Rainwater events are becoming increasingly harsh due to the climate crisis and the warming up of Earth's water bodies. When this is combined with the severity in which developments reduce the natural permeability of sites and increase rainwater runoff, flooding is on the rise and becomes more and more frequent. Increased runoff rates can result in:

•   Erosion and sedimentation of waterways.
•   Water quality declines due to pollutants being carried off by runoff water which can be detrimental to marine life in receiving waters.
•   A strain on municipal stormwater management infrastructure and the subsequent tax dollars and energy required to pump and treat stormwater that runs off our impervious developments.

A Storm Water Management Plan (SWP) often entails strategies for rainwater harvesting, where stormwater is collected and/or reused to cut down water runoff from the site. Stormwater management strategies address both the **quality and quantity** of stormwater. LEED encourages practices that reduce stormwater runoff and protect surface and groundwater quality. This is accomplished by lowering the impervious surface area and adding structural or non-structural features that retain water onsite to encourage infiltration and reuse.

The simplest strategies involve construction of a dry or bioretention pond and grading (sloping) this site to collect rainwater in the pond and have it naturally treated by running through the ground and replenishing the aquifer below. More advanced and complex strategies use rain barrels or cisterns to collect excess rainwater and reserve it for non-drinking purposes in restrooms and irrigation (more about this in the Water Efficiency category next).

1   **Use Low Impact Development and Green Infrastructure**
    -   Capture and treat 95th percentile rainfall events (i.e., 95% of all rainfall events do not exceed the 95th percentile runoff volume)
2   **Natural Land Cover Condition Management**
    -   **Amount to manage** = (Post Development Runoff – Natural Land Cover Condition Runoff)

Essentially, do not increase the rate of rainwater runoff through development based on local historical data.

The **strategies** are divided into two categories:

•   **Green Infrastructure** – Green Infrastructure covers an assortment of water Management practices and technologies that help water infiltrate, evapotranspire, capture, and reuse stormwater to maintain or restore natural hydrologies.
•   **Low Impact Development** – This alternative stormwater management strategy stresses working with onsite natural features to protect water quality by replicating the natural land cover hydrologic regime of watersheds and addressing runoff close to its source (i.e., maintaining vegetated swales and rain gardens and minimizing impervious cover.)

Here, a common theme in LEED may be observed, where the developmental impact on our natural environment is mitigated. Essentially, the site should be left as is, either in its present state or in a better condition than it was found.

A common misconception in the industry is believing LEED credits restrict and limit design ideas and goals, which could not be further from the truth. Take this credit, for example, where instead of constructing a concrete sidewalk, it can be easily be converted into an open grid pavement (50% permeable), as we see below, which achieves the same intention to function as a pathway. Conversely, this credit can be entirely ignored, but the project would, in the process, forfeit 2–3 points. The item that deserves recognition here is that not only does LEED allow a project team to choose the credits they want to pursue, but many of the credits will enable them to select the strategy to be adopted to achieve them. LEED keeps the design open-ended and not restrictive (Figure 5.1).

Figure 5.1 Example of open grid pavement.

LEED Credit – Heat Island Reduction – SSc5 – 1–2 Points

During the summer months, the temperature consistently drops and plummets when traveling from a dense and populated downtown core into the rural countryside environment (Figures 5.2 and 5.3). This is largely attributable to the heat island effect, defined as the thermal difference between developed and undeveloped areas. Urban areas experience higher temperatures in comparison to their surrounding rural counterparts due to a number of reasons, such as:

- The exorbitant amounts of dark hardscapes such as asphalt and concrete that absorb and store heat during the day and release it at night.
- This heat remains in the city as a result of inadequate ventilation arising from narrow streets, high buildings, vehicle exhaust, and other pollutants.
- The lack of evapotranspiration magnifies the heat island effect and contributes to urban smog.

**Little vegetation or evaporation causes cities
to remain warmer than the surrounding countryside**

Figure 5.2 The heat island effect.
Source: Image by rawpixel.com on Freepik.

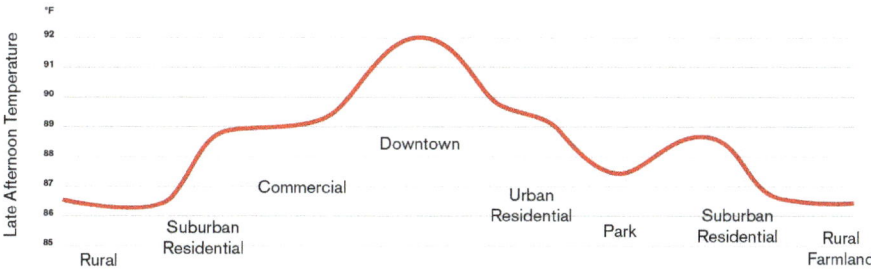

Figure 5.3 Graph to show the thermal difference between developed and underdeveloped areas.

Figure 5.4 Thermograph of a developed area.

The outcome of the heat island effect is warmer temperatures in urban settings during the summer and an energy surge from our building cooling loads, which artificially impacts wildlife habitats (Figure 5.4).

LEED refers to 'Heat islands' as urban areas that experience a hike in temperature from 2 to 10 degrees which are significantly warmer than the rural areas surrounding it, addressing both roof and non-roof components in two different options under this credit. This credit under the SS category encompasses cool pavements and cool roofs that refer to materials with lower **emissivity,** higher **albedo,** and, subsequently, higher **Solar Reflectance Index (SRI).** The **measurement terms** that need to be considered here are:

Emittance or emissivity is the ability of a material to emit and release heat by radiation.

Solar reflectance or albedo is the fraction of the solar energy reflected by a surface determined as a number between 0 and 1. The higher the number, the better the roofing material reflects energy.

Solar reflectance index (SRI) is a composite measure of the constructed surface's ability to reject solar heat or radiation into the atmosphere. It is a measure that combines the value of reflectivity and emissivity. It is used for roof surfaces. The value for a standard black roof is zero (reflectance 0.05, emittance 0.90), while that of a standard white roof is 100 (reflectance 0.80, emittance 0.90).

Solar Reflectance (SR) – It is the fraction of solar energy that is reflected by a non-roof surface on a scale of 0 to 1. Black paint is 0, and white paint is 1.

Strategies for reducing the heat island effect:

1. **Non-roof** or on the ground
   - **Hardscapes**: For the purpose of bringing down the heat island effect, higher solar reflectance materials are to be used, and the areas of impervious hardscape reduced. Employing cool pavements with a three-year aged SR of .28 (or Initial SR of 0.33) or higher or open grid pavements, which are 50% or more permeable, are also beneficial for rainwater management. A sample solution is to construct buildings closer together on a site.
   - **Shading**: Provide shading for 50% of the hardscape area by taking advantage of existing tree canopies **OR** planting new trees, anticipating their size and shade within a period of 10 years of installation.
   OR Place shading structures covered by solar panels **OR** architectural features with an SR of 0.28 or higher.

2 **Roofs**
   Install roofs using one of the following strategies:
   1   Utilize roof materials with the following SRI requirements:

	Slope	*Initial SRI*	*3-year aged SRI*
Low-sloped roof	≤ 2:12	82	64
Steep-sloped roof	> 2:12	39	32

   A low-sloped roof has greater direct exposure to the sun and heats up faster, resulting in a higher SRI requirement.

   1   Construct one of the following types of vegetative roofs:
       a   **Extensive Green Roof** – This vegetative roof is not designed for human access and requires minimal maintenance.
       b   **Intensive Green Roof** – This vegetative roof entails a variety of plants and human uses. Native and adaptive plants are encouraged.

The credit synergies of a green roof can contribute to the following credits:

1   Open Space (intensive)
2   Rainwater management (intensive and extensive)
3   Protect or restore habitat (intensive and extensive)
4   Heat Island Effect Reduction (intensive and extensive)
5   Probable energy reduction, given that the green roof provides a better insulating R-value and holds in heat.

Green roofs are expensive to build and maintain, but time and again, they have proved to be the only option for priceless vegetation in a downtown core.

1   **Parking**: Uncovered outdoor surface-level parking lots are massive contributors to heat island effects, as cars can get scorching hot when parked outside on a sunny summer day, and they are also a destructive and an absolute waste of space. In this case, urban heat can be drastically reduced by:

   a   Locating **75%+ of parking spaces under the building itself, roofs, or shades**.

   b   Building a **parking garage/deck, vegetative roof, or energy generation systems.**

   c   Considering the use of **photovoltaic (solar) panels on top of a parking garage that charge EV charging stations** that can contribute to the following credits:

      i   **Heat Island Effect**

      ii   **Green Vehicles**

      iii   **Renewable Energy**

Figure 5.5a Example of an intensive green roof.

Figure 5.5b Solar panels on top of a parking lot.

This credit may seem counterintuitive and unreasonable to our theme of increasing density (Figures 5.5a and 5.5b). However, by adopting such strategies, both density and mitigation of the heat island effect can be achieved.

LEED Credit – Light Pollution Reduction – SSc6 – 1 Point

Flickr user: Jpstanley "Light pollution: It's not pretty" / CC BY 2.0

Nowadays, no matter how clear the night sky may be in urban areas, the visibility of a vast percentage of stars is simply impossible. Such a situation can be owed to the excessive or intrusive artificial night lights that cause light pollution and annoy our neighbors. This is a grave concern that not only affects humans but also confuses animals as they mistake streetlamps for the moon. LEED promotes well-designed, effective lighting systems that reduce light pollution and resolve all problems relating to it. The light pollution reduction credit is bent on diminishing light trespass from the building and its site, reducing sky glow in urban areas to increase night sky access, improving night-time visibility through glare reduction, and reducing development impact on nocturnal environments. In essence, this credit necessitates any light generated on the site to be restricted to the site. Nevertheless, different areas have different light pollution allowances, as exhibited below as the first step in this credit:

Lighting Zones Created by the Model Lighting Ordinance

Determine the project's lighting zone according to the requirements of IESNA RP-33 as shown below:

2   **LZ0** – No Ambient Lighting
3   **LZ1** – Dark (Park and rural settings)
4   **LZ2** – Low (Residential areas and neighborhood business districts)
5   **LZ3** – Medium (Commercial/Industrial and high-density residential)
6   **LZ4** – High (Major city centers and entertainment districts) (Figure 5.6).

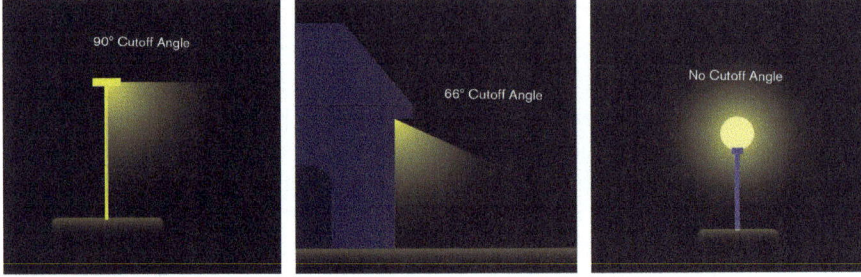

Figure 5.6 Varying the cutoff angle can restrict the amount of horizontal/vertical
       light emitted

There are two options to achieve this credit, and both these options are
concerned with restricting uplight and light trespass off the site.

1   **Backlight-Uplight-Glare (BUG) Method**
     Appertaining to the Illuminating Engineering Society (IES) TM
     standard, the luminaire ratings per lighting zone should be met but not
     exceeded.
2   **Calculation Zone**
     Under no circumstance must the maximum admissible percent-
     age of total luminaire lumens (how bright the light is) emitted above
     horizontal/vertical be surpassed.

Light Pollution Reduction Strategies

• Cover exterior lights and aim them downwards as opposed to no covered orbs directed towards the sky, which results in light pollution and
  energy dissipation. Lights should be targeted on the ground and not
  shining into the sky.

- Utilize exterior signage that complies with the threshold requirements.
- Methodically map out the location of exterior lights to be insulated from the lot line and reduce light trespass into neighboring properties.

Although the LEED Green Associate primarily focuses on the LEED BD+C Rating System and New Construction market adaptation, it is always beneficial to be familiar with the existence of some credits unique to different market adaptations, such as:

Schools Only – Site Master Plan – SSc7

The SSc7 is comparatively a simple credit where four of the six credits jotted below must be achieved:

- **LT Credit: High Priority Site**
- **SS Credit: Site Development Protect or Restore Habitat**
- **SS Credit: Open Space**
- **SS Credit: Rainwater Management**
- **SS Credit: Heat Island Reduction**
- **SS Credit: Light Pollution Reduction**

Schools Only – Joint Use Facilities – SSc8

The school's facilities should be optimally utilized and shared with the general public or specialized services such as a police station or health care center. Alternatively, a building space owned by another organization could be shared and made accessible to students, such as an offsite gymnasium. The provision of direct pedestrian access and toilet access after hours are also a part of the joint-use facilities.

Core and Shell Only – Tenant Design and Construction Guidelines – SSc7

The project team must create an easily understandable and self-explanatory document to ensure tenants can perceive and implement the sustainable features of the core and shell of the building. This document can assist tenants in keeping their space up to a sustainable standard and make it easier to pursue LEED ID+C.

Healthcare Only – Places of Respite and Direct Exterior Access – SSc7 + SSc8

Human health is directly correlated to healthcare facilities, and as a result, the built environment should do as much as possible to foster occupant well-being. These credits deal with mimicking the outdoor environment indoors (places of respite) or creating an accessible outdoor space such as courtyards or terraces (Direct Exterior Access).

Noteworthy Standards

Erosion and Sedimentation Control Plan must comply with the **EPA General Construction Permit or local codes.**

6 Water Efficiency

How Things Were Formerly Done/Problems

Water is one of the most vital renewable resources on earth compared to other natural resources. The typical average American family consumes more than 300 gallons of water daily at home, and roughly 70% of this use occurs indoors.

Research shows that all across the country, outdoor water consumption accounts for 30% of household use. However, this may not be true in the drier parts of the country and more water-intensive landscapes where it can be much higher. https://www.epa.gov/watersense/how-we-use-water

Fifty percent of U.S. lakes and rivers are considered impaired, i.e., they are too contaminated for swimming, fishing, or drinking.[1] According to the United Nations Environment Program, the rate at which we are going, if our present patterns continue, by the year 2025, two out of every three people will live in water-stressed conditions.

DOI: 10.1201/9781003405856-6

How LEED® Addresses It

LEED's water efficiency solution confronts the potable water usage situation within buildings, cooling towers' water usage, the site's landscaping, and proper metering of water usage. All the credits in this category concentrate on efficiency, whereas the prerequisites are immersed in attempting to reduce waste (11 points).

What LEED Requires for the Credits

LEED calls for strategies to be implemented in developments that reduce the amount of water used. A project pursuing the path of LEED must make sure to comply with the prerequisite for outdoor water use reduction of 30%. Additionally, LEED requires outdoor/indoor water reduction and water meters to have a place in the project to examine, monitor, and consequently efficiently reduce excessive exploitation of water. LEED supports the adoption of strategies, procedures, and technologies that cut down the quantity of potable water consumed in buildings while satisfying the requirements of the systems and occupants. Several strategies come at no cost or provide a rapid payback. Water use reduction can ease the burden and pressure on wastewater treatment facilities.

Before metaphorically diving into the water category and rummaging through it, it is vital to be cognizant of some terminology.

Potable water – Potable water refers to water that meets or surpasses EPA's drinking water quality standards and is approved for human consumption by state or local authorities exercising jurisdiction. Every single credit in this category shares common ground, which is to **minimize potable water use**. Be aware of the fact that if time, money, and energy have gone into making water drinkable, it must be conserved for processes that demand such high-quality water.

Graywater (or greywater) – Household wastewater coming from bathroom and laundry sinks, tubs, showers, and washers is known as graywater. The Uniform Plumbing Code (UPC) defines graywater as untreated household wastewater that has not come in contact with toilet waste, while the International Plumbing Code (IPC) defines graywater as wastewater discharged from lavatories, bathtubs, showers, clothes washers, and laundry sinks. The majority of states do not permit kitchen sinks or dishwashers to be included with graywater as it comes into contact with food.

Blackwater – Wastewater generated from toilets and urinals is called blackwater. Most jurisdictions view water from kitchen sinks and dishwashers as blackwater, which cannot be reused.

Process water – Water employed for industrial processes and building systems such as boilers, cooling towers, and chillers is process water.

Stormwater runoff – Stormwater runoff is referred to as runoff water which occurs as a result of precipitation that flows over ground surfaces and conventionally into storm sewers or waterways.

WaterSense – WaterSense is an EPA-sponsored program that advocates and certifies water-efficient products, programs, and practices. How EnergyStar is to appliances, WaterSense is to flush and flow fixtures. WaterSense assists consumers in identifying water-efficient products and programs that conform to WaterSense water efficiency and performance criteria. Nevertheless, WaterSense fixtures must use some amount of water (i.e., Waterless urinals do not comply).

Gallons per flush (gpf) – This metric is the flow rate measurement unit for flush fixtures such as water closets and urinals.

Gallons per minute (gpm) – Gallons per minute is the flow rate measurement unit for flow fixtures such as faucets, showerheads, aerators, and sprinkler heads, which measures how many gallons of water flow out of flow fixtures per minute.

Waterless urinals – These no-flush urinals are designed to work entirely in the absence of water or flush valves by passing urine through a sealing liquid. There are two types of waterless urinals – Cartridge-based and Non-cartridge-based units.

Dual flush – The eco-friendly dual flush water closets minimize water use by allowing one to choose between a full flush for solid waste and a half flush for liquid waste. These toilets can save around 2/3 of the water used for flushes, slashing the water bill and saving money in the long run.

Water Efficiency versus Conservation – While water conservation addresses the policies and activities that manage the use of water, the intention behind efficiency is to minimize the water required for a specific purpose. **Conservation** is defined as the practice of trying to reduce unnecessary usage of water entirely, and **efficiency** emphasizes doing a great deal more with the least amount of water for the same task. From a personal perspective, conservation and efficiency strategies should be stretched to their limit and utilized to the maximum prior to opting for other expensive and energy-intensive alternatives, such as desalination or piping water across the country.

Full-Time Equivalents (FTEs) – The FTE of a project must be consistent across all credits. FTE refers to a regular building occupant who spends 40 hours a week within the project building. Part-time or overtime occupants have FTE values contingent on the hours they spend inside the building per week divided by 40. FTEs are based on a standard 8-hour occupancy period per day. An 8-hour (5 days a week) full-time occupant has an FTE score of 1.0. Being conversant with the steps to compute FTE is of utmost importance, as this is sometimes the only calculation that appears on the Green Associate Exam. FTEs are used to ascertain the amount of water that must be reduced and determine the amount of bicycle parking required to be provided in the location and transportation credit category.

FTE identifies the total number of building occupants based on their occupancy types:

- **Full-time staff**
- **Part-time staff**
- **Peak Transients (students, volunteers, visitors, customers, etc.)**
- **Residents** (Figure 6.1).

WATERSENSE LABELED PRODUCTS

Ever since the first WaterSense labeled toilets hit store in 2007, more and more product types have earned the WaterSense label, and the total number of WaterSense labeled models has continued to grow.

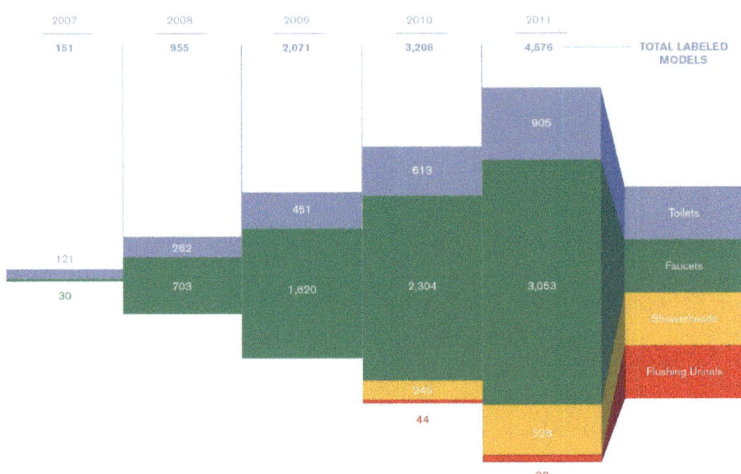

Figure 6.1 Graph to show the increase in watersense labeled products between 2007 and 2011.

In this category, we are introduced to three pairs of prerequisites and credits. It is pretty evident and self-explanatory how the prerequisite requires a certain reduction percentage, and the subsequent credit rewards the project incrementally for further reduction.

Outdoor Water Use Reduction	
Percentage Reduction	*Points (BD + C)*
30%	Prerequisite
50%	1
100%	2

LEED Prerequisite and Credit – Outdoor Water Use Reduction – WEp1 + WEc1 – Prerequisite – 2 Points

In the past, people watered plants with the aid of above-ground sprinkler systems, where a large majority of the water was either lost through evaporation or runoff even before actually reaching the plant's root, its intended destination. This credit cautions against conventional sprinkler systems, as well as irrigation as a whole, for that matter.

Projects pursuing LEED must adhere to the **prerequisite** threshold of 30% for outdoor water use reduction.

Points are incrementally awarded based on the amount of potable water offset by choosing strategies such as:

- Landscaping with native or adaptive plants to reduce or eliminate irrigation demands.
  - It is outlined in the rainwater management section in the chapter on sustainable sites native plants that previously lived on the site can thrive again in similar conditions on or around the building.
- Adopting xeriscaping landscaping (specifically designed for sites where water conservation is practiced and lands that are susceptible to drought) by selecting drought-resistant vegetation that requires little to no irrigation.

Table 6.1 Table to show the points granted to building projects based on the percentage reduction of indoor water use

Indoor Water Use Reduction	
Percentage Reduction	*Points (BD + C)*
20%	Prerequisite
25%	1
30%	2
35%	3
40%	4
45%	5
50%	6
55% (EP)	7

- Mulching to ensure moisture retention and avoid evaporation of the plant's water.
- Reduce or eliminate turf grasses that need to be irrigated constantly.
- Employ efficient irrigation systems such as Drip irrigation systems or micro-misters which can lead to a 90% water use reduction in contrast to the conventional above-ground sprinkler installations.
- Installing demand meters that ensure the irrigation system does not set off during periods of natural rainfall.
- Utilizing non-potable water that is unsafe for consumption for irrigation purposes, such as captured rainwater, graywater, or municipally claimed wastewater (Table 6.1).

LEED Prerequisite and Credit – Indoor Water Use Reduction – WEp2 + WEc2 – Prerequisite – 7 Points

In LEED, there is mention of a certain percentage of improvement. However, when it is a new building we are evaluating, what can it be compared to? Here, the straightforward use of a baseline known as the Energy Policy Act of 1992/2005 (EPAct) is used to define a conventional (old) consumption rate of a version of our building.

The Energy Policy Act of 1992/2005 (EPAct) established water conservation standards for water closets, showerheads, faucets, and other plumbing fixtures as mentioned below:

- Conventional toilets: 1.6 gpf
- Conventional urinals: 1.0 gpf
- Private lavatory (bathroom) faucets: 2.2 gpm
- Public faucets: 0.5 gpm (Private or public distinction is based on location and use)
- Conventional kitchen faucets: 2.2 gpm
- Conventional showerheads: 2.5 gpm

Standards for specific appliances are as follows:

- Residential clothes washer – ENERGY STAR or performance equivalent
- Commercial clothes washer – CEE Tier 3A
- Residential dishwasher – ENERGY STAR or performance equivalent
- Pre-rinse spray valve – consumption ≤1.3 gpm (4.9 lpm)
- Ice machine – ENERGY STAR or performance equivalent, and use either air-cooled or closed-loop cooling

Water use reduction calculations are not based on the number of fixtures but, in turn, on the number of FTEs and the default gender ratio between FTE occupants, which is 1:1. In practice, a LEED online calculation tool stores the EPAct's flush and flow data, and the project team inputs the number of FTEs in the building as well as proof of purchase of water-efficient flush and flow fixtures. Although the prerequisite solely rewards efficient fixtures, the credit rewards non-potable water usage to offset where potable water is conventionally used, such as our toilets.

Flickr user: bunnicula, "dual flush toilet" / CC BY-ND 2.0

Water use reduction strategies:

- Install water-efficient plumbing fixtures such as:

- **Efficient flow fixtures**: Lavatories, sinks, and showerheads with lower gpm rates.
- **Efficient flush type fixtures**: Dual flush toilets, waterless toilets/ urinals, composting toilets, high-efficient toilets HET.

- **Utilize non-potable water** such as captured rainwater, graywater, or municipally claimed wastewater for toilets and urinals.
- **Employ water-efficient fixtures, non-potable water, or municipally treated wastewater**.
- **Treat on-site wastewater** to tertiary standards and reuse it for non-potable applications.

Note – Rule out water reuse from raw naturally occurring water bodies, rivers, groundwater, well water, seawater, and discharge from open-loop geothermal.

Percent Improvement calculations = {(Baseline – Actual) / Baseline)} × 100 as most reduction calculations are done.

Wikimedia Commons User: llanbar6, "Package type cooling tower" / CC BY-SA 4.0

LEED Prerequisite and Credit – Water Metering – Building Level (WEp3)/Submetering (WEc4 – 1 Point)

A critical path to ongoing sustainability is tracking or metering the actual water usage to learn and understand the volume of water utilized over the course of time. Modern water distribution systems have many components in the project and on its site, which results in ample leakage opportunities. Meters are an action-oriented proactive approach that aims to locate potential leaks, fight water loss, and determine unusual usage patterns and improvement opportunities for the largest consumers.

The Building level water metering prerequisite demands that all potable water sources, such as public supply, on-site wells, and on-site potable water treatment systems, be metered as a whole. In the situation where there are multiple potable sources, additional meters may be required. The municipality or utility company will often, following request, give access to their water meter as they use it to charge the building. In any case, if they do not attain a water meter from the authorities, a building can install its own meter downstream to produce comparable results.

This meter must collect data for total water consumed in the building and on its associated grounds, as well as commit to sharing annual summaries with the USGBC for 5 years. Currently, LEED BD+C buildings do not have a recertification requirement, but this new prerequisite insinuates the likelihood of recertification eventually at some point in the future. Please be advised that the only LEED rating system currently requiring recertification is the LEED O+M for existing buildings to portray actual energy usage reduction every 5 years.

The water metering credit requires two or more permanently installed system-level water meters for the following subsystems:

1   Irrigation water (80% of the area)
2   Indoor plumbing fixtures and fittings (80%+)
3   Domestic hot water (80%+)
4   Boilers that use 100,000g+ / year
5   Reclaimed water (100%)
6   Process water (80%+)
    -   Humidification systems
    -   Dishwashers
    -   Clothes washers
    -   Pools

The water metering credit often requires the installation of multiple within complex buildings. Though this can wind up being a costly and cumbersome affair, they can potentially stop leaks quickly.

Wikimedia Commons user: André Karwath, "Water meter (aka)" / CC BY-SA 2.5

LEED Credit – Cooling Tower Water Use – WEc3 – 1 Point

While it is not imperative to understand the principle behind how the cooling towers operate, they are actually pretty straightforward, and the cool water is employed in the building's HVAC system. This is primarily where process water is used, which is basically water used in building equipment. Cooling towers assist with the air conditioning process by cooling water in a closed-loop system through evaporative cooling and heat extraction. The warm water exits the building at the end of the cooling cycle and is sprayed over a large surface area while a large fan cools the water down. However, during every cycle, some water evaporates through evaporative cooling, and unwanted dissolved solids build-up, resulting in the removal of this unwanted concentrated water (blowdown) to prevent inefficient scaling in the equipment. Makeup water must then be added to counterbalance and make up for blowdown and evaporation. This credit strives to maximize the number of times the water can cycle through the system before being discharged by blowdown.

This credit necessitates a one-time potable water analysis to ascertain the concentration of at least five dissolved solids and ensure maximums are not outpaced. For supplementary points, a project can decrease the concentration levels in makeup water to achieve at the very least ten cycles prior to blowdown or use 20% recycled non-potable water (such as rooftop collected rainwater or air conditioning condensate) to accomplish the minimum of ten cycles. Bear in mind that every credit in this category attempts to reduce potable water use in and around the building and its processes (Figure 6.2).

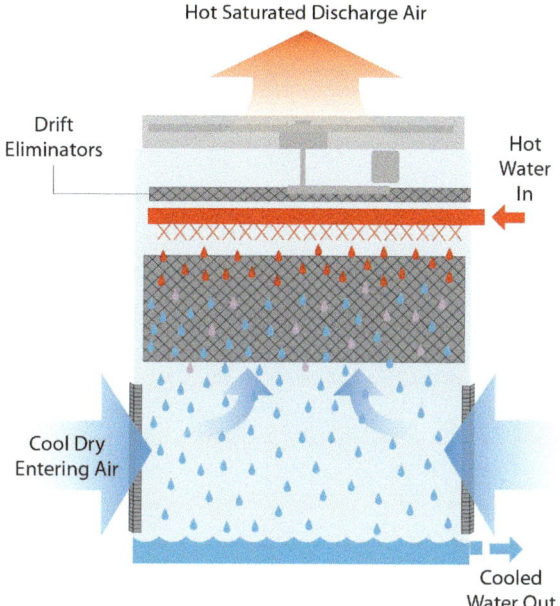

Figure 6.2 Diagram of a cooling water tower.

Noteworthy Standards

• EPAct
• Watersense
• The WaterSense Water Budget Tool automatically derives rainfall and evapotranspiration from the project's zip code for the outdoor water reduction credit.

Note

1 Olivia Rosane, "50% of U.S. Lakes and Rivers Are Too Polluted for Swimming, Fishing or Drinking," *World Economic Forum,* April 5, 2022, https://www.we-forum.org/agenda/2022/04/50-of-u-s-lakes-and-rivers-are-too-polluted-for-swimming-fishing-drinking

7 Energy and Atmosphere

Overview

Way Things Were Done in the Past/Problems

According to the U.S. Department of Energy, electricity generated from fossil fuels such as natural gas, oil, and coal negatively and substantially harms the environment throughout its life cycle.[1]

How LEED® Addresses It

In this chapter, LEED's solution talks about efficient energy and refrigerant usage in buildings. This chapter covers four topics that are explored in detail – Energy demand, energy efficiency, renewable energy, and outgoing energy performance. Energy and Atmosphere is a relevant chapter since it contains the most points one can achieve within the LEED rating system (33 points).

Noteworthy Standards

- ASHRAE 90.1
- Montreal Protocol
- Green-e renewable energy certification
- Renewable Energy Certificates (RECs) or Tradable Renewable Certificates (TRCs)

DOI: 10.1201/9781003405856-7

LEED's Requirements for Credits

To meet the prerequisites and credits in this valuable category, projects are mandated to submit a commissioning plan, energy model, refrigerant usage schedule, and proof of energy metering and renewable energy sourcing. Likewise, demand response programs also call for proof of contracts with local utility providers.

Among all LEED credit categories, it is the Energy and Atmosphere category that yields the most possible points as it has the potential to alleviate both greenhouse gas emissions and operational costs of the building. Deterioration of the environment begins with non-renewable resource extraction and transportation, followed by refining and distribution, and draws to a close with consumption (burning fossil fuels). Buildings in the U.S. consume approximately 39% of the energy and 68% of the electricity produced annually, contributing to an estimated 38% of the global carbon emissions as a consequence of greenhouse gases. The Energy and Atmosphere category advocates efficiently designed buildings and sourcing energy from renewables (Figure 7.1).

- The reference chart above[2] clearly portrays the high energy use intensity in the US. This occurs due to one of the following reasons: Energy is comparatively economical.
- Majority of the population has access to energy.
- Extreme climates with cold winters and hot summers.

Consequently, owing to the above reasons, the population is not afraid to use inexpensive and available energy to fulfill their needs. As previously noted, the initial step that requires our attention in our energy dilemma is cutting down the current energy consumption rates. Such findings also put us in a unique and privileged position as, presently, there is a massive room for improvement and scope for development to catch up to the rest of the world's energy use intensity.

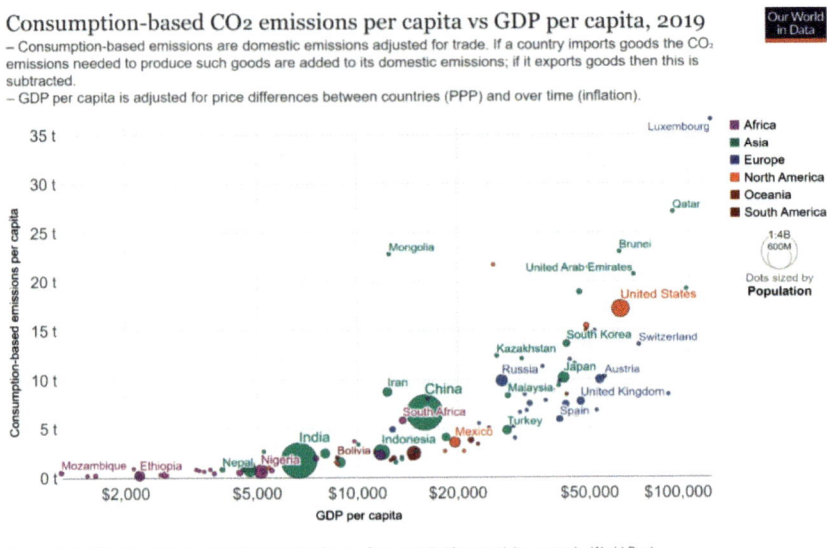

Figure 7.1 Consumption-based CO$_2$ emissions per capita vs GDP per capita in 2019, highlighting high energy use intensity in the US.

This category addresses and brings into focus both energy and refrigerant usage in the building. The crux of the Energy and Atmosphere category copes with the following topics:

- Energy demand
- Energy efficiency
- Renewable energy
- Ongoing energy performance

Much like Water Efficiency, four pairs of prerequisites and credits, along with three individual credits, will be scrutinized and probed into, all geared towards mitigating the impact of energy production for building consumption.

LEED Prerequisite – Fundamental Commissioning and Verification – EAp1

LEED Credit – Enhanced Commissioning and Verification – EAc1 – 2–6 Points

Commissioning is a quality-oriented process employed to verify that the project's energy-related systems are planned, installed, calibrated, executed, and maintained, adhering to the Owner's Project Requirements (OPR), the basis of design (BOD) and construction documents. Commissioning confirms that the building owner derives the best performance out of the efficient systems and proper implementation. While fundamental commissioning of building energy systems is required within a prerequisite, enhanced commissioning is addressed in a credit.

The underlying notion behind this credit stems from the owner paying a premium for high-efficiency and often unique building equipment with the desire to lower the operating and maintenance costs over the life of the building. However, in spite of everything going as per plan, occasionally, there may occur situations where a designer or subcontractor may not possess the adequate experience required for working with these systems. Hence, a commissioning agent familiar with the system intervenes and assumes the role of a third party to verify the equipment will operate at its design case efficiency. However, the commissioning agent must have similar-sized projects and report directly to the owner to provide unbiased recommendations and evaluations.

Benefits of Commissioning

- Reduced energy consumption
- Lower operating costs
- Fewer contractor callbacks
- Better building documentation
- Enhanced occupant productivity
- Verifies that the commissioned systems perform in accordance with the Owner's Project Requirements.

Commissioning Process Activities

- Designate an individual as the commissioning authority (CxA) to lead, review, and oversee the completion of the commissioning process.
- Document the Owner's Project Requirements (OPR).

- Formulate the basis of design (BOD) to include the OPR.
- The CxA must review and assess OPR and BOD to verify that owner's requirements are included.
- Develop and incorporate commissioning requirements into the construction documents.
- Develop and implement a commissioning plan.
- Verify the installation and performance of the systems to be commissioned.
- Complete a summary commissioning report.

Mandatory Commissioned Systems (MEP) – Eap1

In the mandatory section of commissioning, the commissioning agent must Verify and Document that the following systems and assemblies are planned, designed, installed, tested, and maintained as per the OPR (Figures 7.2):

- **Mechanical** – HVAC&R systems
- **Electrical** – Lighting and daylighting controls
- **Plumbing** – Domestic hot water systems (DHW)
- **Renewable energy systems**

Figure 7.2 The four mandatory commissioned systems: mechanical, electrical, plumbing, and renewable energy systems.

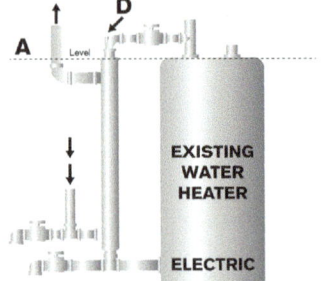

Enhanced Commissioning – EAc1

In an effort to earn points in the Enhanced Commissioning credit, a project team may navigate through either one of the following compliance paths:

Enhanced Systems Commissioning – Path 1 (3 points):
- Review contractor submittals.
- Verify inclusion of systems manual, occupant training requirements in construction documents, and seasonal testing.
- Review building operations 10 months following substantial completion.
- Develop an ongoing commissioning plan.

Path 2 (4 points):
- Achieve path 1 AND
- Include procedures in the Commissioning plan
- AND/OR -

Envelope Commissioning (2 points):
- Follow EAp1's requirements as they apply to the building envelope, which is often a location of energy loss if the structure is not adequately sealed.

LEED Prerequisite – Minimum Energy Performance – EAp2

LEED Credit – Optimize Energy Performance – EAc2 – 1–18 Points

This credit functions in a similar fashion to the water reduction credit in water efficiency but substitutes the word 'water' with 'energy consumption'. This credit offers the maximum potential point count to add to the building's LEED scorecard, where project teams can prove to be the most creative. The credit rewards active energy-saving solutions from high-efficiency HVAC and lighting to passing solutions such as increased insulation and triple-glazed fenestrations (windows and doors). Take a moment to think about all the potential credit interactions, including synergies and tradeoffs.

LEED discusses the building's rational utilization of energy through a prerequisite that sets limits, establishing a minimum level of energy efficiency for the proposed building to limit environmental and economic impacts associated with inefficient energy consumption. Optimized energy performance and higher percentages of energy efficiency are necessary to score points. The number of points earned depends on the percentage of energy cost savings, and the energy performance of a building counts on the reduction of energy demand by the building and the efficiency of the building systems as compared to a baseline, in this case, ASHRAE 90.1-2010.

ASHRAE stands for the American Society of Heating, Refrigerating, and Air-Conditioning Engineers, which focuses on building systems, energy efficiency, indoor air quality, refrigeration, and sustainability within the industry. As observed in preceding chapters, LEED takes advantage of existing proven standards fabricated by industry experts to add value to the rating system as a whole.

It is highly probable for at least one ASHRAE standard to be on the LEED Green Associate Exam, and here focus is on Appendix G of ASHRAE Standard 90.1-2010, serving as our baseline for energy consumption.

LEED offers three options to achieve the energy performance-related credit/prerequisite:

Option 1 – Whole Building Energy Simulation (1–18 Points)

Energy modeling is adopted, with a view to analyzing the improvement of a new building performance, where it is compared to a theoretical baseline (reference building) that meets Appendix G of ASHRAE Standard 90.1-2010. Energy models cover all energy end uses, including lighting, HVAC, and domestic hot water, as well as process energy, including office equipment, computers, elevators, escalators, kitchen cooking and refrigeration, laundry washing, and drying machines. Energy modeling is a large and booming industry, and as the years go by, the list of its uses grows, and more municipalities begin to require energy simulations to obtain building permits. Various building energy modeling software such as EE4, eQuest, IES, and EnergyPlus is used to carry out building performance simulation for analyzing the energy flow within specified parameters within buildings.

Energy simulations allow the project team to make better decisions based on the results of the model's analysis. First, multiple building designs are drawn in a short time frame to discover a design that optimizes the building's annual energy consumption while meeting the goals of the building without ever touching construction equipment. The model takes into account all mechanical and passive building features, including its location, climactic data, and architectural orientation. The final design reports are submitted via LEED online to show percentage improvements (Table 7.1).

As evident in the energy modeling report, there exist two different design results yielding two quantities of energy reduction over the ASHRAE 90.1-2010 baseline. In package 2, it can be noticed that an 18%+ reduction results in 7 points total. In this option, if a project has onsite renewable energy generation (such as Solar PV), the amount of energy generated can be applied to this credit by subtracting it from our design package's consumption resulting in a lower energy consumption total. It should be noted that points per credit will not be tested on the LEED Green Associate Exam as they are reserved for the LEED AP+ Exams (Table 7.2).

Table 7.1 Example of an energy modeling report comparing the annual energy consumption of two different designs

End Use	Modeled Annual Energy Consumption (GJ)		
	Baseline Design	*Package 1*	*Package 2*
Lighting	1,994	1,994	1,994
Misc. Equip.	928	928	928
Heating	10,907	9,977	6,881
Cooling	469	436	461
Heat Rejection	723	723	723
Pumps	520	517	509
Fans	1,733	1,099	1,099
Domestic HW	3,048	2,119	2,119
Total	**20,322**	**17,792**	**14,714**
% Reduction compared to ASHRAE 90.1-2010	**0**	**1.0%**	**18.1%**

Table 7.2 Table to show the points granted to building projects based on the percentage reduction of energy use

Energy Performance

Percentage reduction	*Points (BD+C – New Construction Only)*
5%	Prerequisite
6%	1
8%	2
10%	3
12%	4
14%	5
16%	6
18%	7

(Countinued)

Energy Performance

Percentage reduction	Points (BD+C – New Construction Only)
20%	8
22%	9
24%	10
26%	11
29%	12
32%	13
25%	14
38%	15
42%	16
46%	17
50%	18
54% (EP)	19

In rare scenarios, creating an energy model may be overly complicated or impractical, or perhaps the owner may not want to pay the $12,000 on the average associated fee. In such cases, the two prescriptive path options operate like a checklist and for every box checked, a point is achieved to a much lower possible max score.

Option 2 – Prescriptive Compliance: ASHRAE 50% Advanced Energy Design Guide (1–6 Points)

For instance, in a Medium to Large Box Retail Building:

- Building envelope, opaque: Roofs, walls, floors, slabs, doors, and vestibules (1 point)
- Building envelope, glazing: fenestration – all orientations (1 point)
- Interior lighting, excluding lighting power density for the sales floor (1 point)
- Additional interior lighting for the sales floor (1 point)
- Exterior lighting (1 point)
- Plug loads, including equipment choices and controls (1 point)

Option 3 – (Prerequisite Only) – Prescriptive Compliance: Advanced Buildings™ Core Performance™ Guide

This option looks at Section 3: Enhanced Performance Strategies from ASHRAE 90.1 and requires compliance with the following chapters:

3.5 Supply Air Temperature Reset (VAV)
3.9 Premium Economizer Performance
3.10 Variable Speed Control
To be eligible for Option 3, the project must be less than 100,000 sq ft.

A More Comprehensive View of the Energy in Our Buildings and Tools to Achieve Reduction

Energy Demand

Energy savings are computed based on the demand for energy that a building can reduce compared to a baseline energy model. During the Charrette, the Owner Program Requirements (OPR) are established, and the project team must set goals regarding the project's energy demand and the energy savings they achieve.

EPA's Target Finder: Target Finder is EPA's online tool that allows the project team to set an energy target and then compare their project's design energy and cost to a pre-determined goal. The target would then be measured against a variety of design strategies. The program then outputs the amount of energy the average building in the area (by type and zip code) and a high-performance building in the same area would consume. Target Finder is mandatory compliance for the **LEED BD+C for Schools prerequisite**.

Energy Star Portfolio Manager: EPA's Energy Star Portfolio Manager is an online interactive energy management tool that tracks and assesses energy and water consumption as well as GHG emissions across an entire portfolio of buildings in a secure online environment. It utilizes consumption data and cost of operations to track 100+ metrics and is instrumental in the LEED O+M for the existing building's rating system.

Strategies for reducing energy demand

- **Energy goals**: Define energy-saving targets at an early stage in the project and verify their achievement.
- **Building size**: A larger facility demands more significant amounts of energy and more resources. Hence, the building area should equal and correspond precisely to what is needed to meet its future occupant's needs.
- **Building orientation**: A well-oriented building can benefit from natural ventilation, solar energy, passive heating, and daylighting. Computer-simulated 3D models help designers and architects predict how the building will perform prior to being built and designed accordingly.
- **Building envelope**: A high-performance building envelope insulates itself against heating and cooling losses.
- **Energy monitoring**: Feedback and energy monitoring systems assist occupants in recognizing and reducing the building's energy demand.
- **Building Systems**: Ensure that the Heating, ventilation, and air conditioning (HVAC) and DHW (domestic hot water) systems are appropriately sized and operating at their rated efficiency (Figure 7.3).

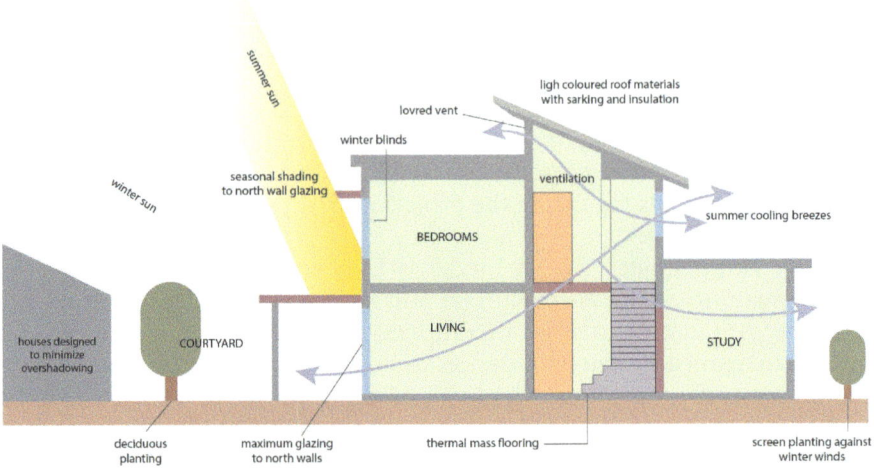

Figure 7.3 Examples of design strategies for reducing energy demand in the home.

Energy Efficiency

Strategies implemented to reduce energy demand are the reason for the increasing energy efficiency.

Energy efficiency strategies

- **Passive design and thermal energy storage:** Determine the best building orientation and massing to benefit from the passive design opportunities such as daylighting, natural ventilation, and passive heating and cooling from the sun and the wind (Figure 7.4).
- **High-performance building envelope:** Installing high-performance glazing systems and efficiently insulating the building envelope will prevent heating and cooling losses and provide thermal comfort for building occupants. The less treated (heat/cool) air that escapes translates into less untreated outdoor air that needs to be heated or cooled.
- **High-performance building systems:** Select good-quality, energy-efficient HVAC, plumbing, electrical, and lighting systems, and perform life cycle analysis to evaluate their overall impact.
- **Verify and monitor performance:** Confirm that the systems are installed as per owner requirements and, subsequently, monitor their performance after occupancy to ensure that the building systems are functioning as designed.

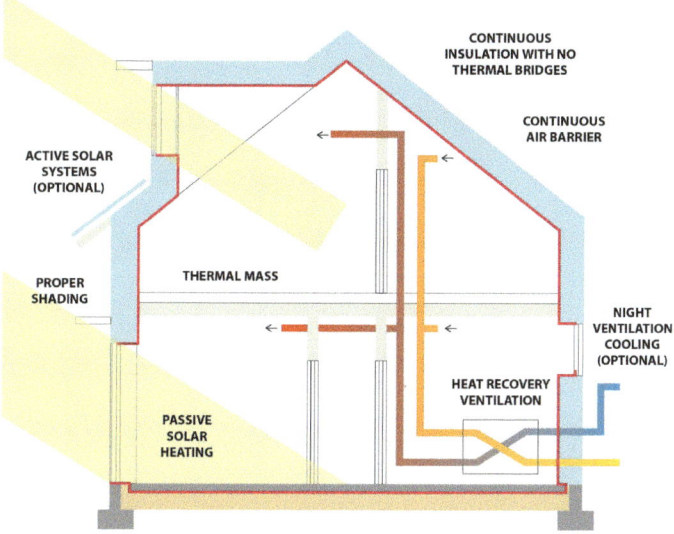

Figure 7.4 Examples of passive design and thermal energy storage to reduce energy demand.

LEED Prerequisite – Building-Level Energy Metering – EAp3

LEED Credit – Advanced Energy Metering – EAc2 – 1 Point

Arguably energy metering is typically one of the most commonly overlooked attributes of sustainable building design, as you cannot measure what you do not meter (Figure 7.5). A project can solely evaluate a building's performance through metering and adopting corrective action when a system is not functioning near its design case efficiency. Similar to how buildings must meter water consumption, energy metering must also be carried out. In fact, it is all the more necessary as energy is the highest-priced aspect of a building's operations. Metering can help identify opportunities for additional energy savings on a building system-level scale.

The **Prerequisite** necessitates permanently installed meters that can aggregately provide total building energy consumption, which can be compiled into monthly/annual summaries and shared with the USGBC for 5 years. This task may come at no cost to the building, provided the utility company grants access to its energy meter.

The **Credit** requires a more advanced system for all individual energy end uses that represent 10% or more of the total annual consumption of the building. The most common consumers of 10%+ energy of the building

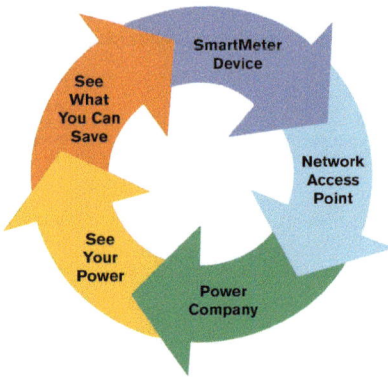

Figure 7.5 You cannot measure what you do not meter – the benefits of installing a
SmartMeter device.

include heating, cooling, lighting, and sometimes, plug loads. Throughout
the past decade, owing to hardware becoming less expensive due to the
Internet of Things revolution, accurate smart meters, which are easy to install, have become commonplace in the market.

The energy meters must:

- Record at a minimum of hour intervals
- Record Demand and Consumption
- Transmit data to Remote Location

Building Automation System (BAS) is software used to control, regulate, and
monitor building energy demand and consumption. The automation system
maintains occupant comfort while meeting energy consumption goals and
identifying mechanical, electrical, and plumbing system woes.

LEED Prerequisite – Fundamental Refrigerant Management – EAp4

LEED Credit – Enhanced Refrigerant Management –
EAc6 – 1 Point

Refrigerants are fluids used to transfer thermal energy in heat pumps, air
conditioning, and refrigerating systems (Figure 7.6). Chlorofluorocarbons

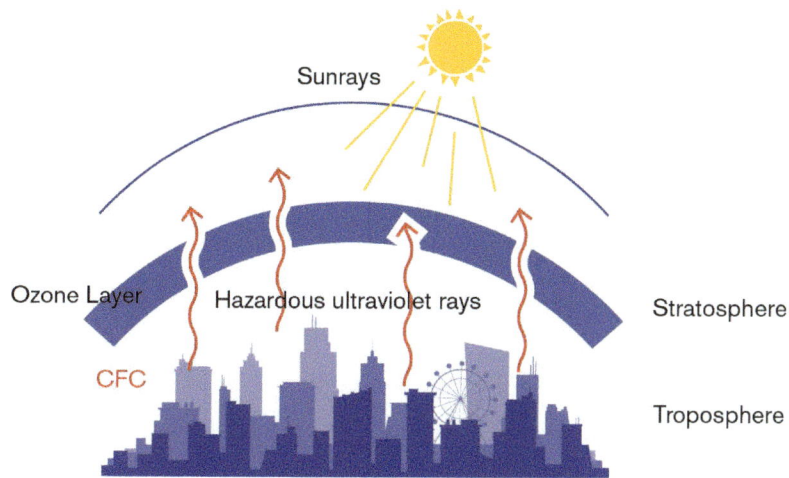

Figure 7.6 Chlorofluorocarbons (CFCs) cause damage to the ozone layer, reducing its ability to absorb UV radiation from the sun.

(CFCs) used in refrigerant equipment cause damage to the ozone layer, thereby reducing its ability to absorb the sun's ultraviolet radiation. The most common example of a CFC is called Freon. The **Montreal Protocol,** signed by many nations in 1987, bans the production of **CFCs** and has established a phase-out date for the use of hydrochlorofluorocarbons (**HCFCs**) in 2030. CFCs and HCFCs are categorized as ozone-depleting substances because of their high Ozone Depletion Potential (**ODP**). The dilemma when selecting refrigerants is that refrigerants with low ODP tend to have high Global Warming Potential (**GWP**), such as Hydrofluorocarbons (HFC). Therefore, the tradeoff among efficiency, ODP, and GWP must be analyzed when choosing refrigerant equipment. **GWP is a metric that examines each greenhouse gas's ability to trap heat in the atmosphere compared to carbon dioxide**

Refrigerants include:

- ChloroFluoroCarbons (CFC) (Figure 7.7)
- HydroChloroFluoroCarbons (HCFC)
- HydroFluoroCarbons (HFC)
- Halocarbons – Primarily used in fire suppression systems

Natural Refrigerants

- Carbon Dioxide (CO_2)
- Water (H_2O)
- Ammonia (NH_3)
- Hydrocarbons (HC)
- Air

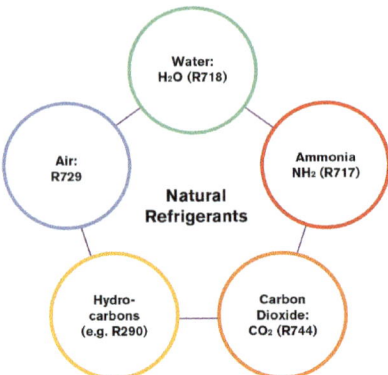

Figure 7.7 Examples of natural refrigerants.

LEED's Fundamental Refrigerant Management prerequisite requires zero use of CFC-based refrigerants in the new building's HVAC & R systems, whereas the Enhanced credit requires ODP+GWP to be under a specified value. The prerequisite is generally automatically achieved in countries that have signed the Montreal Protocol of 1987, banning it in new buildings since 2010.

Strategies to achieve this credit/prerequisite:

- Install HVAC&R systems using no refrigerants.
- Install HVAC&R systems using non-CFC refrigerants.
- Choose fire suppression systems and equipment that use no CFCs, HCFCs, or halons.
- If an existing building is undergoing a major renovation and currently uses CFCs, the building must phase out the CFCs within 5 years and reduce annual leakage to 5% or less.

Please Refer to Appendix (1) for a Required Reading on Refrigerants

LEED Credit – Renewable Energy Production – EAc5 – 1–3 Points

Now that the four pairs of prerequisites and credits have been reviewed and probed into, next in line are the three standalone credits in the Energy and Atmosphere category, known as **Renewable Energy Production.** So far, the concentrated focus has been on reducing energy consumption. However, it is also possible to make a dent in fossil fuel energy sources by generating onsite renewable energy. The energy that is naturally replenished and derived from natural resources is called renewable energy. This energy is more sustainable and environmentally sound than fossil fuels owing to the fact that there is no finite amount of it. LEED distinguishes between renewable energy that is produced onsite and offsite power purchased from renewable sources. This credit deals with onsite renewable energy.

Energy generated from renewable sources can reduce carbon emissions and offer local environmental benefits by reducing air pollution. Generating onsite renewable energy decreases the building's reliance on market energy prices and reduces energy lost through transmitting it via long distances, as the sooner energy is used, the more efficient it is. Thus, energy generated onsite is either utilized by the building itself or sold back to the grid to meet the closest demand nearby, known as net metering. For this very reason, points are calculated based on the equivalent cost of energy reduction over the whole building's annual energy cost as determined by our energy model in the previous credit for optimizing energy performance. In the case where the energy model is not completed, the building's energy cost is determined by the US Department of Energy's Commercial Buildings Energy Consumption Survey database (Prescriptive path).

Renewable energy sources include:

- Wind energy
- Solar thermal, active, and passive
- Photovoltaic
- Biofuels
- Geothermal heating
- Low impact hydroelectric
- Wave and tidal energy

LEED Credit – Demand Response – EAc6 – 1–2 Points

As in the case of commodities, energy is most exorbitant when there is constrained supply and excessive demand for it, such as during a hot summer afternoon when outdoor temperatures are peaking. The graph below illustrates the correlation between the cost of energy usage and time. At peak demand, energy providers must resort to what is known as a 'Peaker plant' which can quickly meet society's energy demand but are also expensive to operate, inefficient, and produce many harmful emissions. In this case, a thought that comes to mind is, "Why not just generate energy overnight during off-peak hours and store it for use at peak times?" Well, though that thought process is logical and correct, the current energy storage technology is still being developed to meet our financial and roundabout efficiency goals. In the meantime, emphasis should be given to collectively reducing the building's energy consumption during those peak hours by enrolling in a demand response program (Figure 7.8).

A demand response program enables the energy provider (utility company) to cut energy from demand response participants during peak hours. The utility company and building owners enter into an agreement that during those peak hours when energy generation prices are at an all-time high, the utility company will pay the building owner to shed a minimum of 10% of the building's energy demand. On receiving an incoming signal from the utility, called a demand response event, the BAS must react by shedding 10%+ of its energy demand automatically. This is attainable by increasing the building's global temperature by 1–2 degrees, turning off decorative features, or dimming lights. Demand response is described as low-hanging fruit and has the potential to avoid the construction of additional energy production facilities. It is counterintuitive as a demand response program has the energy provider paying the consumer to actually use less of the product they are selling (energy). This doubles down on the theme as even energy producers understand how much less expensive it is to reduce energy use rather than generate more.

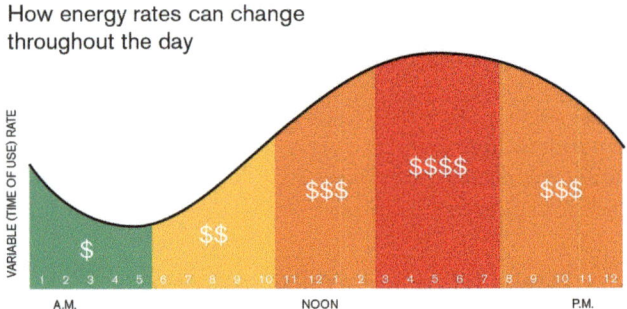

Figure 7.8 Energy rates throughout the day.

Green-e

LEED Credit – Green Power and Carbon Offsets – EAc7 – 1–2 Points

Earlier in this chapter, points for generating onsite renewable energy were covered in detail. Now, it is time to shed light on using offsite generated energy from a renewable supply. There are two ways to meet this credit's requirement – directly using offsite renewable energy in the building or indirectly using renewable energy through the purchase of RECs.

It is to be noted here that the project must purchase 50%+ of the building's electricity consumption to achieve this credit. The percentage of the green power purchased is based on the quantity of **energy consumption**, not energy costs. Strategies to earn this credit are as follows:

- This credit is based on the quantity of energy consumed.
- Purchase power from a **green e-certified provider** on closed or open electricity market to reduce Scope 2 emissions (from electricity, see below).
- Purchase RECs or TRCs representing the 1 MWh electricity generated from a renewable source.
- The qualified resources must have come online since January 1, 2005, for a minimum of 5 years to be delivered at least annually.
- Alternatively, a project can purchase carbon offsets to reduce scope 1 emissions, as one carbon offset equals one carbon dioxide equivalent of greenhouse gases.

Scope 1 emissions – Direct emissions – Natural gas burned on site (i.e., Boiler or furnace) – mitigated by carbon offsets.

Scope 2 emissions – Indirect emissions – Greenhouse gas (i.e., Purchased electricity) – mitigated by green power or RECs.

Green-e is a certification program for renewable energy. LEED typically recognizes renewable energy if it has been certified by the **Center for Resource Solution** or meets Green-e's requirements.

RECs, or TRCs, represent electricity produced from renewable energy sources sold separately from the commodity. RECs are a tradable

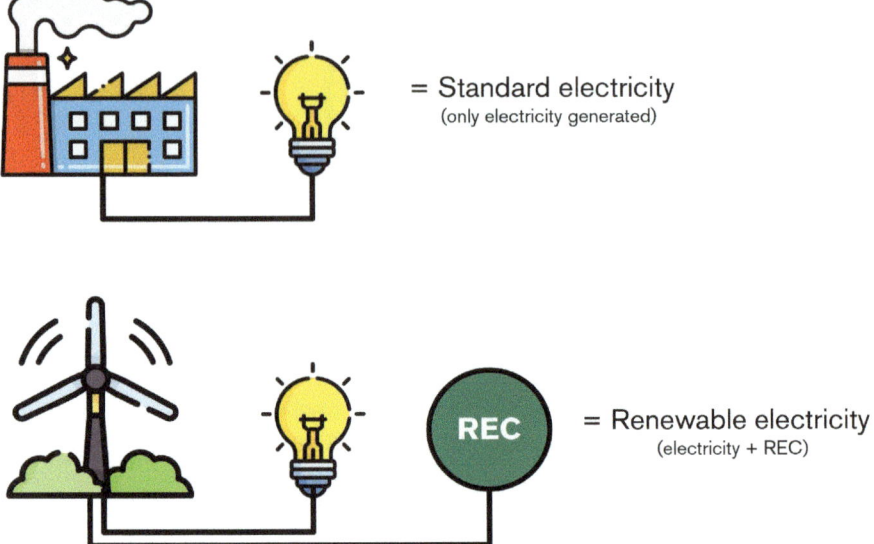

Figure 7.9 Renewable energy certificates represent one megawatt-hour (MWh) of electricity generated from an eligible renewable energy resource.
Source: Icons created by Freepik.com - Flaticon

commodity, and one REC represents one megawatt-hour (MWh) of electricity generated from an eligible renewable energy resource. In the case where there is no green-e-certified power supplier within feasible proximity to your project, REC essentially purchases renewable power for a building that would otherwise be using standard electricity. RECs accomplish the ultimate goal of reducing fossil-fuel-sourced electricity, although indirectly (Figure 7.9),

Noteworthy Standards

ASHRAE 90.1
GREEN-E CERTIFICATION

Notes

1 United Nations Economic Commission for Europe, "Life Cycle Assessment of Electricity Generation Operations," 2021, https://unece.org/sites/default/files/2021-10/LCA-2.pdf
2 Max Roser, "The World's Energy Problem," *Our World in Data,* December 10, 2022, https://ourworldindata.org/worlds-energy-problem

8   Materials and Resources

How Things Were Done in the Past/Problems

Construction and demolition debris account for 40% of our total solid residual waste stream.[1]

How LEED® Addresses It

Material selection and material disposal are the two key issues reviewed in this chapter. In the interest of getting a more accurate picture and deeper understanding of this credit category, sustainable materials and strategies to reduce waste sent to landfills have been tackled. Alongside LEED's solution to encourage and promote the selection of sustainable materials and the reduction of waste produced from the construction, operation, and demolition of buildings are subjects of discussion. (13 points).

What LEED Requires for the Credits

LEED necessitates the prerequisites to be addressed before getting to the credits. In this case, the prerequisites include an easily accessible, dedicated area for collecting and storing recycling materials that must be made available to the entire building. As for the credits, LEED requires proper recycling

DOI: 10.1201/9781003405856-8

and/or salvage of non-hazardous construction and demolition debris by developing and implementing a **Construction Waste Management Plan**.

Image by freepik - www.freepik.com

Terms to understand before proceeding any further:

Life Cycle Impacts: This includes raw-material extraction and processing through transportation, product use, and ultimately the disposal of materials composing it. LEED promotes and advocates reusing or recycling a product at the end of its operational lifespan versus disposing of it, and this approach is known as the cradle to cradle. Likewise, LEED wants to serve as a market mover and ensure all products undergo a Life Cycle Assessment (LCA).

Purchasing Policies: LEED encourages the utilization of sustainable materials through this category. It determines the types of materials that should be considered and qualify for a high-performance building and ascertain how to minimize the life-cycle impacts of these materials. Purchasing policies should explicitly specify where the items are sourced during the construction and operation of the building. Furthermore, third-party certification of sustainable products is critically essential in material selection. Note that LEED does not certify products but instead supports organizations that certify environmentally responsible products (Ex. FSC certified wood).

Waste reduction: The preferred waste management strategies described by the U.S. Environmental Protection Agency (EPA) are **source reduction, reuse, recycling, and recovery**. Material reuse or recycling are forms of waste diversion that help prevent materials from winding up in landfills or incineration plants. The **EPA** defines the following **waste management strategies**:

- **Source Reduction**: Source reduction refers to the practice of designing, manufacturing, purchasing, or employing materials in ways that facilitate lowering the amount of unnecessary or excessive materials in the building.

- **Reuse**: This strategy stops waste at its source as it delays, hinders, or completely eliminates that entry into the waste collection and disposal system.
- **Recycling**: Recycling is the process of converting materials that would otherwise be disposed of as waste into valuable resources.
- **Recovery:** This technique involves using waste materials for a useful purpose, either through proper recycling and composting materials or waste-to-energy plants such as incineration.

Credit metrics and cost calculations: Some of the following credits are based on material area, weight, or cost. A material baseline is established using the actual material cost OR 45% of the total construction costs, including labor and equipment.

Please pay particularly close attention to this credit category as it is the one that has altered the most since LEED version 3, and most examinees are seen struggling with it.

LEED Prerequisite – Storage and Collection of Recyclables – MRp1 – Mandatory

This prerequisite lessens the amount of landfill-bound waste matter generated by the building occupants throughout the entire life or duration of the building. The easier it is to recycle, the better the odds and likelihood for building occupants to follow through with recycling. For this purpose, **a designated space with** an easily **accessi**ble and dedicated area must be equipped to the entire building to collect and store recycling materials. The recycling systems can either be commingled, where the recyclables are gathered together in a single bin, or separated based on the type of recyclable into different individual bins on site. Materials to be recycled must include, **at the very least, paper, corrugated cardboard, glass, plastics, and metals. Furthermore, the project must also offer areas to recycle hazardous materials such as batteries, electronics, or mercury-containing lamps.**

LEED Prerequisite – Construction and Demolition Waste Management Planning – MRp2 – Mandatory

LEED Credit – Construction and Demolition Waste Management – MRc5 – 1–2 Points

As mentioned in the introduction, whether constructing new buildings, renovating, or demolishing and replacing old ones, a great deal of waste is produced at the end of the day, ending up in landfills or dumping grounds.

This prerequisite and credit seek to scale down construction and demolition waste discarded in dumping grounds and incineration facilities by recovering, reusing, and recycling materials. This credit is required to recycle and/or salvage non-hazardous construction and demolition debris by developing and implementing a **Construction Waste Management Plan. While the prerequisite strictly mandates a plan without any minimum diversion rate, the credit requires a minimum of 50% of waste to be diverted from the landfill** (Table 8.1).

The Construction Waste Management Plan must:

1   Identify and set waste diversion goals for **a minimum of five materials** (structural and non-structural).
2   Specify whether the materials will be **separated or commingled** as well as the destination where they will eventually be recycled or donated.
3   Provide major **waste streams and diversion rates.**

Table 8.1 Table to show the points granted to building projects based on the percentage of construction waste diverted

Construction Waste	
Percentage Diverted	*Points (BD+C)*
Plan	**Prerequisite**
50% + 3 streams	1
75% + 4 streams	2

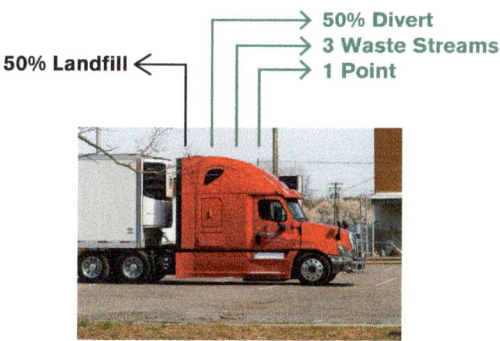

Figure 8.1a Building projects which divert a minimum of 50% of waste to be diverted from the landfill can earn 1 point from LEED (BD+C).

Waste diversion rates are based on volume/weight salvaged/recycled over total waste produced (Figure 8.1a). These calculations exclude excavated soil, land-clearing debris, and alternative daily cover (ADC). Waste-to-energy systems may be included in some instances where the credit cannot be met or fulfilled.

There are two options to earn points:

Option 1 – Diversion into multiple waste streams
Calculated by weight or volume as a percentage of total waste
Option 2 – Reduction of Total waste material (Source Reduction)
Maximum of 2.5 pounds of construction waste/square foot of buildings
    floor area

LEED Credit – Building Life-Cycle Impact Reduction – MRc1 – 2–5 Points

No matter whether a brand new building is constructed or an existing one is reused on a previously developed site, there is always room to reduce the amount of waste hauled to landfills and virgin materials purchased for the build. However, reusing parts of an existing building while attempting to meet the current project's goals may prove to be challenging and tricky, making this one of the toughest credits to attain. This credit is achievable regardless of the situation, provided the right team expertise and collaboration are available from the beginning of the project.

This credit awards points for **adaptive reuse** of existing building resources (provided it is a previously developed site) or for reducing materials use through life cycle analysis. Adaptive reuse is the process of renovating a space for a purpose different from the original.

There are four options for credit compliance:

Option 1 – Historic Building (5 points)

Previously, in the Location and Transportation credit category, historic areas were discussed. However, a historically designated building itself is called on in this credit. The building or historic district must be listed or eligible for listing in the local, state, or national register of historical places to qualify for this credit. Often, the governing body enforces the law for maintaining a historic building façade, and this credit directs to maintain the existing building structure, envelope, and interior non-structural elements of a historic building or contributing building in a historic district.

Option 2 – Abandoned or Blighted Building (5 points)

Abandoned or blighted buildings (deteriorated and unsafe) negatively impact the neighborhood they are located in, reduce the property value of buildings within the area, and threaten the stability of such communities. However, it is neither obligatory nor necessary to demolish them entirely, as parts of the building's structure may remain intact and can be reused. For this option, the credit asks to:

- Maintain 50% of the **surface area** of the existing building
  - 25% of the damaged area may be excluded from the credit calculation

Option 3 – Building and Material Reuse (2–4 points)

This option presents an opportunity to reuse or salvage building materials offsite or onsite as a percentage of the **surface area**. Potentially, a building can salvage or restore doors or ceiling systems from existing buildings already onsite or a building being demolished nearby in the vicinity. The option is designed with the objective of collectively minimizing energy and waste associated with construction and demolition.

This credit excludes window assemblies and hazardous materials (the reuse of hazardous materials is completely ruled out). It also excludes materials which contribute towards MRc2, as discussed later on in the chapter.

Option 4 – Building Life-Cycle Impact Reduction – New Construction (3 points)

This option is brand new in LEED version 4 and has the potential to allow brand-new buildings to earn points in this credit. The option requires a life-cycle assessment of the project's structure and enclosure.

- Model a baseline building (using the ISO 14044 standard) and compare them with the proposed design case results.
- The credit requires the life-cycle assessment to demonstrate a minimum of 10% environmental reduction in three of the following **impact categories**:

1 **Global warming potential** (greenhouse gases) in CO2e;
2 **Depletion of the stratospheric ozone layer** in kg CFC-11;
3 **Acidification** of land and water sources in moles H+ or kg SO_2;
4 **Eutrophication** in kg nitrogen or kg phosphate;
5 **Formation of tropospheric ozone** in kg NOx or kg ethene; and
6 **Depletion of non-renewable energy resources** in MJ.

Building Product Disclosure and Optimization – MRc2 to MRc4

While this category mainly aims to minimize energy and waste using sustainable buildings and materials, the remaining portion of the category wanders off the theme and highlights how the materials used in the buildings are actually created as well as their life cycle impact. As discussed earlier, LEED certifies buildings and utilizes dozens of smaller standards to certify products. The decision is totally up to the developers and builders to continue to demand sustainably made materials, as manufacturers and suppliers who ignore market demand will not thrive and prosper.

The following credits work with LCA and go the extra mile to promote and support responsible material sourcing and avoid the inclusion of harmful and poisonous chemicals in construction products. Building Product Disclosure and Optimization grant points for selecting and disclosing the material ingredients or contents used to manufacture the product with transparency.

The following credits are based on the cost of materials as a percentage of the total material cost (actual total cost of materials excluding labor or, if unknown, the default denominator is 45% of total construction costs). These credits strictly deal with permanent building materials, and in the event that the project team includes optional products and materials, such as furniture and MEP items, the actual value of those items is to be added to the default value for all other products and materials to stay consistent.

Location Valuation Factor for the Building Product Disclosure and Optimization Credits (MRc2 to MRc4):

A significant factor that has a hand in the emergence of transportation emissions is when a raw material sourced in one region of the world gets shipped out to another region to be manufactured and, in due course, gets dispatched to its final destination to be sold and used. The intent of this factor is to enhance the value of locally produced materials which curbs transportation emissions and supports the local economy. These local materials are worth 200% of their cost for credit calculation purposes. Essentially this factor makes it easier to achieve the subsequent three credits by sourcing locally extracted, manufactured, and purchased products by doubling their value in credit calculations to follow suit.

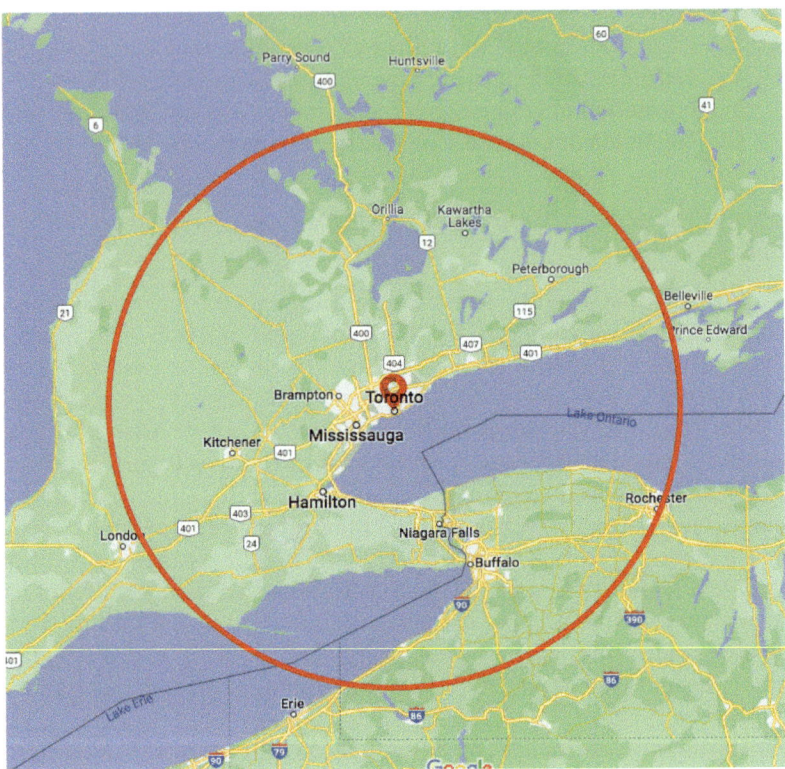

Figure 8.1b 100 mile radius around Toronto.

To comply with this credit, products sourced must be:

1   Extracted, AND
2   Manufactured, AND
3   Purchased,

within 100 miles (160 km) of the project site. The distance must be measured as the crow flies, not by actual travel distance (Figure 8.1b).

LEED Credit – Building Product Disclosure and Optimization – Environmental Product Declarations (EPDs) – MRc2 – 1–2 Points

Environmental Product Declarations (EPDs) reward teams for singling out manufactured products that have been verified and confirmed as having improved life cycle impacts. EPDs prioritize transparency, document a product's ability to mitigate the effects of global warming, ozone depletion, and air/water pollution, and promote corporate social responsibility. Compliance with different standards results in different multiples of the product's cost (0.25–1). Points can be earned by satisfying one of the two options jotted below:

Option 1 – Environmental Product Declarations

The product scope analyzed through EPDs must cover the product's life cycle from cradle to gate [from resource extraction (cradle) to the factory gate]. This credit requires the use of a minimum of 20 different permanently installed products sourced from at least five different manufacturers that meet referenced standards and weightings as displayed in the chart below:

OPTION 1 – EPD (1 point)		
Reference Standard	*EPD Type*	*Valuation factor*
ISO 1405,14040,14044+ 21930 OR EN 15804	Product specific Type III EPD	1.0
ISO 14025,14040,14044+ 21930 OR EN 15804	Industry Wide (Generic) EPD	0.5
ISO 14044	Publically Available, critically reviewed life-cycle assessment	0.25

Basically, EPDs are a standardized way [through the International Organization for Standardization (ISO) standards] of conveying credible information about the environmental effects and implications associated with a product or system's raw material extraction, energy use, chemical makeup, waste generation, and emissions to air, soil, and water.

Option 2 – Multi-Attribute Optimization

This option ensures that 50% (by cost) of materials [or only 25% (by cost), if locally sourced and thus, doubled in value] are third-party certified products demonstrating a reduction in 3+ of the Building Life-Cycle Impact Reduction factors as cited below (see option 4 of MRc1).

1 **Global warming potential** (greenhouse gases);
2 Depletion of the **stratospheric ozone layer**;
3 **Acidification** of land and water sources;
4 **Eutrophication**;
5 Formation of **tropospheric ozone**; and
6 **Depletion of non-renewable energy**

In this credit, LEED offers project teams the option of achieving the credit either through a specific number of products (option 1) or based on their value as a percentage of the overall cost of materials (option 2).

LEED Credit – Building Product Disclosure and Optimization – Sourcing of Raw Materials – MRc3 – 1–2 Points

Though in the previous credit, the cradle to gate impact of a product's manufacturing process was analyzed, in this credit, scrutiny is on whether the raw material was extracted or sourced in a responsible manner. For instance, did the wood originate from a clear-cut or selectively cut forest, or was the mine strip-mined or remediated after resource extraction? Instead of EPDs, here, the emphasis and reliance are on corporate sustainability reports (CSRs) based on widely recognized frameworks and standards, which can give an overview and shed light on product supply chains and identify sources of raw material extraction.

This credit encourages the adoption of products and materials that readily possess life cycle information and environmentally, economically, and socially preferable life-cycle impacts.

Terms to be familiar with in this credit are:

- **Pre-consumer**: Pre-consumer materials are excess/damaged materials re-used for a variety of purposes, such as an error in a factory – ISO 14021.
- **Post-consumer**: Post-consumer materials are made from recycled materials that served their intended use in the past and are now being reused, such as a milk carton recycled in a blue bin – ISO 14021.
- **Bio-based product**: These products are derived wholly or partly from plants, animals, or marine products (animal skin or hide is not rewarded).
- **Chain of Custody**: Chain of Custody is a process that tracks a product right from extraction to end use (FSC or SFI wood certification).
- **CSRs**: Corporate Sustainability Reports are verified based on recognized standards or third parties.

Once again, to demonstrate compliance, there are two options, one based on the number of materials and one on the cost of certified materials. They are as follows:

Option 1 – Raw material source and extraction reporting

Ensure the selected products possess self-declared or third-party verified corporate social responsibility reports that show a commitment to long-term ecologically responsible land use and are bound to reduce environmental harm from extraction and/or manufacturing processes. Under this option, 20 products attained from 5 manufacturers must adhere to the following standards as displayed in the table below.

OPTION 1 – Reporting

Reference Standard	Reporting Type	Valuation Factor
GRI,OECD,UC Global Compact, ISO 26000	Third Party verified CSR	1.0
Internal	Self-declared raw material supplier report	0.5

Standards include:
- Global Reporting Initiative (GRI) Sustainability Report
- Organisation for Economic Co-operation and Development (OECD) Guidelines for Multinational Enterprises
- UN Global Compact: Communication of Progress
- ISO 26000: 2010 Guidance on Social Responsibility
- USGBC-approved program: Other USGBC-approved programs meet the CSR criteria.

CSRReports

Option 2 – Leadership Extraction Practices

Use products that fulfill at least one of the responsible extraction criteria below for at least 25% (by cost) of the total value of permanently installed building products in the project:

1 **Extended producer responsibility** – An example of EPR is a company that participates in a buy-back program of their product from the consumer at the end of its useful life when it is no longer of any use to them, such as a major battery.
2 **Bio-based materials** – Bio-based products must satisfy the Sustainable Agriculture Network's Sustainable Agriculture Standard. However, they rule out hide products, like leather and other similar animal skin material.
3 **Wood products – LEED piggybacks off of two wood certifications**. Wood and wood products are certified in accordance with the Forest Stewardship Council (FSC) or Sustainable Forestry Initiative (SFI) standard for a chain of custody certification.
4 **Materials reuse** – Materials reuse encompasses salvaged, refurbished, or reused products. If product values are counted in this credit, they cannot be double-counted in the MRc1 credit, as noted earlier.
5 **Recycled content** – Post-consumer recycled content criteria are valued at 100% of their cost, while on the other hand, pre-consumer content criteria only count towards 50% of their cost as those materials were errors rather than used for their intended purpose.

In short, this credit seeks to increase transparency in mining, quarrying, agriculture, forestry, and other branches of industry, which can sometimes fly under the sustainable radar.

LEED Credit – Building Product Disclosure and Optimization – Material Ingredients – MRc4 – 1–2 Points

This credit category examines and analyzes permanently installed products in buildings, with a view to ascertain what precisely the products used in day-to-day lives are composed of. Under this credit, LEED incentivizes a team to select products with transparently reported ingredients and have verified improved life cycle impact by mitigating harmful substance use. Chemical ingredients and components inventoried have to resort to a LEED-accepted standard or method to minimize the use and generation of toxic substances. This credit comprises three options with diverse approaches to report or optimize material ingredients.

Option 1 – Material Ingredient Reporting (1 Point)

In this option, the emphasis is on understanding the chemicals used to manufacture products that surround our everyday life. With the objective of achieving this credit, a project must ensure to use of a minimum of 20 different permanently installed products sourced from at least five different manufacturers that list 99%+ of a product's ingredients through the following manufacturer programs.

Health Product
DECLARATION

Option 2 – Ingredient Optimization (1 Point)

Unlike option 1, option 2 not only focuses on understanding the product's chemical makeup of the products but also ensures they do not contain certain specific chemicals or substances. Attaining this credit is possible, if the project utilizes products that document their material ingredient optimization using the paths below for at least 25% (by cost) of the total value of permanently installed products in the project. A product must comply with the below-mentioned standards:

1   GreenScreen v1.2 Benchmark
2   Cradle to Cradle Certified
3   International Alternative Compliance Path – REACH Optimization
4   USGBC-approved program

Option 3 – Supply Chain Optimization (1 Point)

The third option goes one step further and interprets the entire supply chain of a product. This is becoming more popular and frequent in the ever-growing ESG (Environmental, social, and governance) industry. For the purpose of achieving it, a project must use building products for at least 25% (by cost) of the total value of permanently installed products in the project that document at least 99% (by weight) of ingredients for materials and **supply chain processes**.

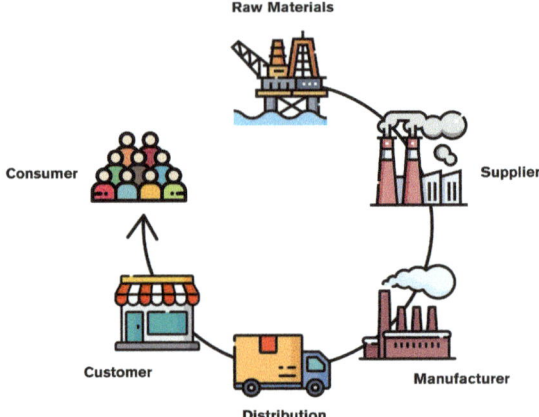

Figure 8.2 A typical supply chain.
Source: This image has been designed using resources from Flaticon.com

Third-Party verified supply chain must address the following:

- Chemical hazard/exposure information,
- Health risks
- Optimization of risk chemicals
- Safety and stewardship of chemical ingredients
- Communication processes for evaluated chemical safety

While credits outside of the LEED BD+C Rating System and subsequent New Construction market adaptation are rarely examined, it is imperative to view additional examples and their basis in other specialized building types, such as hospitals (Figure 8.2). For instance:

MRcHealthCare – Furniture and Medical Furnishings

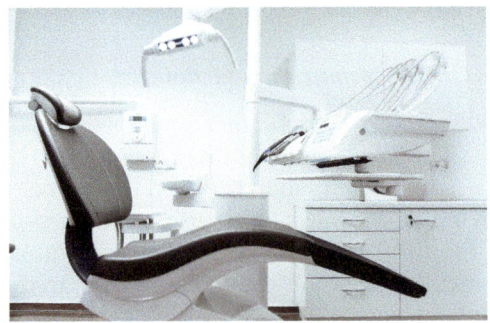

The goal of this credit is to enhance and promote the environmental and human health status by ensuring materials comply with one of the three options:

1 Minimal chemical content
2 Testing and modeling of chemical content
3 Multi-attribute assessment of products

MRcHealthCare – Design for Flexibility

Design for Flexibility is a credit that is extensively applied in the construction and management of buildings. It aims to conserve resources associated with them, thereby increasing building flexibility and ease of future adaptive use and the service life of components and assemblies.

This credit encourages the employment of **Interstitial space** [essentially an extra floor (6-9' tall) between floors to allow for easy and effortless renovation], modular equipment, and future-proofing for expansion is encouraged (Figure 8.3).

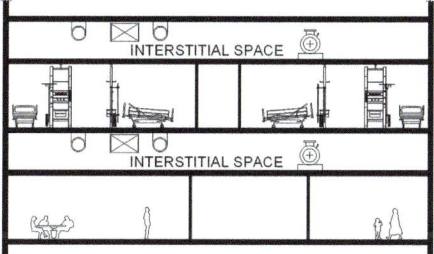

Figure 8.3 Illustration of interstitial space.
Source: Wikimedia Commons User: Ckr5000,
"Interstitial space" / CC BY-SA 4.0

Flickr user: Photo Phiend, "Interstitial space of the dome of the US Capitol, Washington, DC" / CC BY-NC-ND 2.0

MRcHealthCare – PBT Source Reduction (Healthcare Only)

Healthcare has a prerequisite and two credits which are molded and geared towards alleviating PBTs (Persistent Bioaccumulative Toxins) in waste generated through all stages of a product.

On the one hand, the prerequisite draws attention to establishing maximum levels of mercury content in lamps, while on the flip side, its corresponding credit necessitates the project to specify only low or no mercury-contaminated illuminants.

A supplementary credit demands projects to abate or completely eliminate the use of lead, cadmium, and copper.

Noteworthy Standards

- ISO 14044
- Environmental Product Declarations
- GreenScreen v1.2 Benchmark
- Cradle to Cradle Certified
- International Alternative Compliance Path – REACH

Note

1 "Land, Waste, and Cleanup Topics," EPA United States Environmental Protection Agency, Last modified December 8, 2022, epa.gov/osw/conserve/rrr/imr/cdm/pubs/cd-meas.pdf

9 Indoor Environmental Quality

- Indoor air quality (IAQ), note the difference between IAQ and EQ (Environmental Quality),
- Thermal comfort,
- Lighting comfort, and
- Acoustic performance.

Way Things Were Done in the Past/Problem

Indoor Environmental Quality is paramount because, according to EPA, Americans spend at least 90% of their time indoors. The concentration of indoor pollutants is 2–5 times higher than that of outdoor pollution levels. Poor indoor environmental quality adversely affects the health of occupants, and they are prone to illnesses such as asthma and sick building syndrome (SBS).[1]

How LEED® Addresses It

Sustaining a high indoor environmental quality requires collaboration between the building owner, design team, contractors, building operators,

DOI: 10.1201/9781003405856-9

and occupants. LEED makes it a point to focus on improving indoor air quality, enhancing occupant comfort, and promoting well-being. This chapter explores four core issues and the potential solutions to resolve them (16 points).

- Indoor air quality (IAQ), note the difference between IAQ and EQ,
- Thermal comfort,
- Lighting comfort, and
- Acoustic performance.

What LEED Requires for the Credits

Most of the prerequisites and credits in this category emphasize improving indoor air quality. The two key strategies in these credits/prerequisites depend on increasing ventilation rates or reducing contaminants such as tobacco smoke, carbon dioxide, volatile organic compounds (VOCs), and particulates.

Indoor Air Quality (IAQ)

As mentioned in the introduction, the first four credits in this category focus on the breathability of the air we breathe inside buildings. During the pandemic, many individuals who never paid attention previously to the amount of fresh air versus stale air in our built environment became more aware of it. This led to a shift in how our buildings were designed to ensure fresh air ventilation minimums and saw many building codes adopt new requirements.

Let's think of the value of employees in a building versus the building's operating costs. When the salaries and benefits of all employees in our office building were summed up and divided by the square footage of the entire building, humans were paid nearly 100 times more than the annual energy bill per square foot of a building. As crude as it sounds, we went through a commissioning process to ensure our building equipment worked efficiently. So, why not apply the same idea to the humans inside the building who are more valuable? We want our workers to demonstrate the highest quality of work and productivity if they are paid so much.

LEED Prerequisite – Minimum Indoor Air Quality Performance – EQp1 – Mandatory

There is a reason we call fresh air, well, fresh! In most cases, outdoor air is always less contaminated when compared to the indoor air recirculated inside a building. As and when fresh air is pumped into the building, contaminated

**Conventional
Construction**

**Tate's Building
Technology Platform**

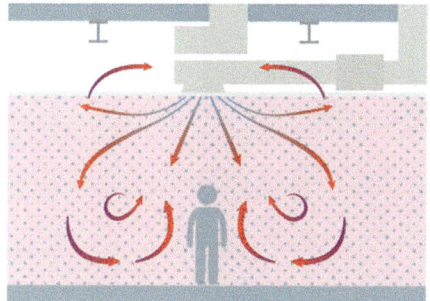

Poor Ventilation
Clean air from the ceiling mixes
with the warm pollutant-filled
air close to the ceiling before
getting to the occupied zone

Improved Air Quality
Fresh air supplied from the floor
improves ventilation as natural
convection moves warmer
stagnant air toward the ceiling

Figure 9.1 Ventilation in a conventional construction and a building using design
to improve air quality.

air is exhausted and replaced. This prerequisite sets minimum fresh air ventilation rates for our buildings and refers to ASHRAE 62.1-2010 as a standard that sets minimum rates for mechanical (HVAC) or naturally (passively) ventilated spaces (Figure 9.1).

We have mentioned the importance of ASHRAE in LEED in previous chapters. This prerequisite uses ASHRAE 62.1-2010 to set fresh air ventilation rates required by the different modes of ventilation based on size and occupant count per space.

Ventilation systems vary from one project to another and include:

- **Mechanical ventilation (active)** – ASHRAE 62.1-2010 determines minimum ventilation rates for various applications.
- **Natural ventilation (passive)** – ASHRAE 62.1-2010 specifies requirements for the size and location of ventilation openings.
- **Mixed-mode ventilation (active + passive)** – ASHRAE 62.1-2010 determines minimum ventilation rates, and any calculation methodology can be used to meet it.

The naturally ventilated spaces must be maintained in accordance with the Chartered Institution of Building Services Engineers (**CIBSE**) applications manual, and an analytical model may be used to confirm that the room-by-room airflows meet the minimum rates required by **ASHRAE 62.1-209.**

LEED Prerequisite Environmental Tobacco Smoke (ETS) Control – EQp2 – Mandatory

Though the ill effects of smoking are widely known, many people continue to smoke cigarettes. Environmental tobacco smoke (ETS) induced by cigarettes, pipes, or cigars poses health hazards such as lung disease, cancer, and heart disease. The purpose behind this prerequisite is to prevent the exposure of building occupants to ETS, also known as second-hand smoke.

To achieve this prerequisite, a project must:

- **Prohibit smoking** inside the building and on-premises within 25 feet of operable windows, entries, and air intakes. Essentially this prevents contaminated air from coming into the building.
- **Post signage** to communicate the smoking policy within 10 feet of all building entrances, which notes no smoking within 25 feet of the entryway.
- **Residential projects** – It may not always be possible for buildings to control the behavior of all their occupants. Hence, in residential buildings, smoking must be forbidden in all common areas as well as within 25 feet of all entries, windows, and air intakes. Each of the units in the residential building must be compartmentalized by:
  - Weather-stripping all exterior and hallway doors and windows.
  - Demonstrating a maximum leakage of 0.23 cubic feet per minute per square foot at 50 Pa of the enclosure.
    - This lowers the likelihood of ETS leaving an individual's unit and coming into contact with other's lungs.
- **Schools** – 'No smoking' signage must be posted at the property line indicating the no-smoking policy within the school property.

LEED Prerequisite – Minimum Acoustical Performance – Schools ONLY – EQp3 +

LEED Credit – LEED Prerequisite – Minimum Acoustical Performance – EQc9 – 1 Point (All Rating systems)

Canteen vector created by pch.vector

Acoustics is an integral element of indoor environmental quality. When teaching in a classroom, or any other space, it is crucial that all students can clearly hear both their instructor and peers. Without this opportunity, a learning environment truly becomes compromised. For this very reason, all LEED for Schools projects must have a minimum requirement for acoustics, while all other building types have the option of pursuing the credit. LEED wants to minimize distractions by limiting noise from the HVAC system, canceling noise from outside the classroom, and reducing internal echo.

To accomplish the prerequisite and credit, spaces must:

1   Achieve a maximum background noise from the HVAC of:
   • 40 decibels (dBA) in schools
   • Other spaces must follow the 2011 ASHRAE Handbook, HVAC Applications, Chapter 48, Table 1.
2   Follow guidelines to minimize reverberation times by coating surfaces with materials or supplies that absorb sound instead of bouncing it back.
   • This can be accomplished by positioning noise-reducing panels along the rear walls and ceilings, helping achieve high-quality sound.
3   Schools only – Utilize strategies to minimize exterior noise, such as barriers placed between the noise source and the classroom.
4   Other rating systems – Meet LEED's sound transmission class (STC) ratings with the aid of sound-dampening insulation placed inside interior walls to prevent sound transfer between rooms.

Not only do quiet environments allow students to learn better, but it also enables employees to work more efficiently. Lately, projects have begun utilizing white noise generators in ceilings or floors to stop sound from traveling across offices, especially in the case of open-concept offices.

LEED Credit – Enhanced Indoor Air Quality Strategies – EQcl – 1–2 Points

The Indoor environmental quality category aims to ensure people breathe clean air, resulting in happy, comfortable, and productive experiences. This LEED credit extends beyond the first prerequisite to ensure a higher quality of indoor air. It protects the occupants from potentially hazardous particulates and chemical pollutants and promotes strategies that minimize human contact with airborne chemicals and suspended particles. This credit focuses on reducing chemicals entering the building while also lessening those being produced inside the building through two options:

Option 1. Enhanced indoor air quality strategies (1 point)

1　**Entryway Systems**

　　Installing 10-foot-long grates or mats, cleaned weekly, helps to capture dirt and debris being tracked into the building by shoes (Figure 9.2).

2　**Interior Cross-contamination Prevention**

　　Sufficiently exhaust contaminated spaces to capture and remove pollutants within confined spaces. For example, a copy/printing room being stuffy, LEED requires an exhaust fan to be fitted in the room, as commonly seen in bathrooms.

3　**Filtration**

　　LEED uses a standard called MERV (Minimum Efficiency Reporting Value) to define the strength of filters used to clean fresh air and free them of airborne particles ventilated into the building. Likewise, LEED also requires MERV 13+ filters to be in accordance with ASHRAE Standard 52.2–2007.

Figure 9.2 Examples of factors affecting air quality in the home.
Source: House vector created by freepik

Option 2 – Additional Enhanced IAQ Strategies (1 point)

1   **Comply with Option 1 above and select one of the following**:
  a   Utilize **air models** to minimize and control the entry of pollutants into the building.
  b   **Increased ventilation** – Increased ventilation ensures bringing in 30% more fresh air than mandated in the LEED prerequisite Minimum Indoor Air Quality Performance as defined by ASHRAE 62.1-2010.
2   **Carbon dioxide monitoring** for densely occupied spaces (25 people or more per 1,000 sq ft) involves mounting CO_2 Sensors that alert the building automation system when the CO_2 concentration levels exceed the setpoint by more than 10%.
3   **Additional Source Control and Monitoring**
    When new materials enter a building, there must be an action plan in place to minimize and flush out the entry of new contaminants into the building. An alarm must be installed as well to indicate unsafe conditions.

Each of the above-mentioned strategies works alone. However, when coupled with other strategies, LEED ensures the highest quality of air inside buildings.

LEED Credit – Low Emitting Materials – EQc2 – (1–3 Points)

This credit highlights the largest source of indoor pollutants, off-gassing contaminants, and chemicals. VOC, which stands for volatile organic compounds, is one term everyone should be aware of. Though VOCs occur naturally, they are found in high concentrations in multiple manufactured materials and supplies used throughout buildings. These materials offgas into the air we breathe. Inhalation of VOCs for extended periods causes several short and long-term chronic health issues, ranging from a runny, itchy nose to cancer.

Similar to the materials and resources category, it is our responsibility to demand low or no-VOC supplies for our building. This applies to even the smallest supplier if they intend to grow in the industry, and they must meet the sustainable needs of the market.

In LEED v4, this credit has been simplified. Tests are run for each layer of a surface, even those not directly exposed to air in the interior space, as defined as everything within the waterproofing membrane. This credit has requirements for seven different categories of materials and two calculation-based methods to show compliance.

1 – Product Category Calculations (you should be familiar with the standards and corresponding category for the LEED Green Associate Exam).

The essential standards include the following:

1   California Department of Public Health (CDPH) Standard Method
2   South Coast Air Quality Management District (SCAQMD)
3   American National Standards Institute (ANSI)
4   Business and Institutional Furniture Manufacturers Association (BIFMA)

Onsite products must meet the following VOC content requirements:

- Interior paints and coatings applied on site – CPDH and SCAQMD 1113
- Interior adhesives and sealants applied on site – CPDH and SCAQMD 1168
- Flooring – CPDH
- Composite wood – California Air Resources Board (CARB) + No added Formaldehyde resins

- Ceilings, walls, thermal, and acoustic insulation – CPDH
- Furniture (include in calculations if part of the scope of work) – ANSI/BIFMA
- Healthcare and School Projects only: Exterior applied products – CARB 2007 + SCAQMD 1168

2 – Budget Calculation Method

If certain products in a category do not meet the criteria, in that case, a team can assess each layer of assembly – paints, coating, adhesives, and sealants, by utilizing the Budget Calculation Method to earn points.

LEED Credit – Construction Indoor Air Quality Management – EQc3 – 1 Point

Recall the first prerequisite in sustainable sites, which addressed the potential for contamination around the project site during construction. Well, if exposed to the inside of a construction site, encountering many more of these potential contaminants is confirmed. Construction and demolition practices introduce pollutants to the building's indoor environment that can have devastating effects over the lifetime of the building. With a view to address problems that originate from construction activities, an IAQ management plan should be developed and implemented during construction and prior to occupancy. LEED award points for managing the project's IAQ at multiple stages of the project. This credit promotes best practices to ensure high IAQ during construction as outlined in the **Sheet Metal and Air Conditioning National Contractors Association (SMACNA) Indoor Air Quality Guidelines, which include:**

- **Protect absorptive materials** installed or stored onsite from moisture to prevent mold.
- If permanently installed air handlers are used during construction, a filtration media with a Minimum Efficiency Reporting Value (**MERV**)

of eight must be provided to all return grills as determined by **ASHRAE 52.2-1999** to prevent contamination of HVAC ducts used throughout the project.

- Before occupancy, all filtration media must be substituted with the final design filtration media.
- Strictly prohibiting the consumption of tobacco products inside the building and within 25 feet during construction.
- Maintaining a clean and organized construction site through proper scheduling, housekeeping, and pathway interruption.

Documentation of construction-based credits such as this is completed by illustrating a plan and displaying pictures from an unbiased third party, such as the LEED consultant, confirms the plan was executed. If the plan is not fulfilled for some reason, the third party must communicate with the contractor and show the issue was remediated. This method applies to all LEED credits completed during the project's construction phase.

LEED Credit – Indoor Air Quality Assessment – EQc4 – 1–2 Points

This credit analyzes the point in time between construction completion and occupancy. Theoretically, this is when the building's air should be least contaminated, as occupants have yet to bring their own forms of contamination into the building. There are two options to achieve this credit (Figure 9.3):

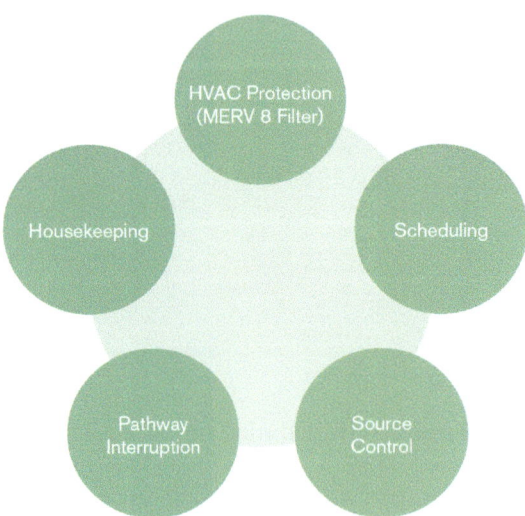

Figure 9.3 Examples of best practice to ensure high indoor air quality during construction.

Option 1 – Flush Out – Air flushing replaces all the existing indoor air with fresh outdoor air prior to occupancy. Projects must supply a total air volume of **14,000** cubic feet of outdoor air per square foot of floor area while maintaining an indoor air temperature of at least **60°F** and relative humidity of not higher than **60%**. Essentially, pumping fresh air throughout the building ensures all construction contaminants have been exhausted, keeping the indoor air odor-free and clean.

Option 2 – Air Testing – On several occasions, a complete building flushout may be prohibited for large buildings as it requires ample amounts of time to exhaust airborne contaminants within the building. Alternatively, air samples are collected during normal occupied hours from areas between **3 and 6 feet above floor level (our breathing zone)** to demonstrate that the maximum contaminant concentrations allowed have not exceeded specifically for **Formaldehyde, ozone, VOCs, and Carbon monoxide (CO).**

Both options portrayed above ensure that the air is as clean as possible and free from impurities before occupancy.

LEED Credit – Thermal Comfort – EQc5 – 1 Point

Thus far, in the Indoor Environmental Quality Credit Category, the sole focus has been on the breathability of the air within a building. However, now it is time to shift the focus on the overall comfort of the indoor environment, beginning with thermal comfort. People are often heard complaining about being either too cold or hot in a space, which most certainly affects their comfort, productivity, and well-being. The main drivers that affect thermal comfort, as described in **ASHRAE 55-2010,** are:

• Occupant activity
• Occupant clothing
• Air temperature
• Radiant temperature
• Air speed
• Humidity

In this credit, besides ensuring that thermal comfort is achievable, building occupants are furnished with the ability to regulate the temperature inside their space, as everyone has their own unique comfort setpoints. This credit has two mandatory options to achieve thermal comfort:

Option 1 – Thermal Comfort Design – HVAC systems and the building envelope must comply with **ASHRAE 55-2010**. Essentially a building's design can offer high thermal comfort potential when accompanied by a high-efficiency HVAC system and well-insulated windows and walls, as per ASHRAE 55-2010.

Option 2 – Thermal Comfort Control – Thermal Comfort Control is defined as the ability to adjust air/radiant temperature, air velocity, or humidity. The most common modes of control come in the form of thermostats and operable windows, and they ensure:

- 50% of individual occupant space has thermal comfort controls.
- 100% of multi-occupant spaces, such as classrooms, have thermal comfort controls.

Wikimedia Commons user: Panek, "Particle counter aerotrak 1" / CC BY-SA 4.

LEED Credit – Interior Lighting – EQc6 – 1–2 Points

Similar to how in the previous credit, thermal comfort and control in a building were explored to maximize occupant comfort, the same holds true about lighting quality and control. If a space is illuminated with either too bright or too dim lights with no controls, our happiness level and productivity will significantly decline. Conventional buildings often provide excessive direct lighting that reflects off surfaces into our eyes, lights that burn out quickly, or uneven lighting throughout a space. This credit attempts to meet occupant lighting requirements through effective design of lighting types, controls, and locations, which includes two options (Figure 9.4):

Option 1 – Lighting Control
- 90% of individual occupant spaces have lighting comfort controls with three modes (on/off/midlevel). The most convenient way to achieve this is by equipping task lights at each desk or just by providing a dimmer switch in each room in residential buildings.
- 100% of multi-occupant spaces have lighting comfort controls with three modes (on/off/midlevel).
  - Lighting for presentations/projections must be separately controlled and located in the same space, and the operator must have a direct line of sight without any obstruction blocking their view.

Figure 9.4 Thermal comfort design ensures airflow can be adjusted to suit individual needs.
Source: Teamwork illustration vector created by pch.vector

Option 2 – Lighting Quality – This option must meet at least 4 out of the 8 strategies listed below:

a) Minimize objectionable glare.
b) Approximate natural light through artificial illumination (bulbs).
b) Maximize lamp life by using LEDs or longer-life fluorescents.
c) Alleviate direct-only overhead lighting.
d) Specify high-reflecting surfaces to illuminate a space with fewer bulbs for:
    a) ceilings + floors
    b) walls
    c) furniture
h) Maintain a ratio of illuminance between objects to ensure a comfortable level of light.

LEED Credit – Daylight – EQc7 – 1–3 Points

When a space is comfortably lit with natural light from the sun, and there is no need to flip on the light switch, one often feels happier, more comfortable, and subsequently more productive. Introducing a proper daylighting design can lead to a decrease in energy consumption and healthy indoor environmental quality.

While that is a potential synergy, the potential tradeoff must also be considered, as windows are always viewed as a worse insulator than a basic wall. Modern construction often makes use of floor-to-ceiling curtain wall systems since glass is a cheap and light building material that is easy to install. However, energy modeling software must be utilized to select the size and location of our glazing (windows) and test the various design options that enable daylighting without compromising the integrity of the building envelope's ability to insulate. This concept is the basis for building science and a starting point for sustainable design.

Shading devices, light shelves, courtyards, atriums, and window glazing are strategies for maximizing daylighting. This credit requires mounting manual or automatic glare-control devices (blinds or curtains) for all regularly occupied spaces through one of the three options discussed below.

The first two options demonstrate daylighting through computer simulations, while the third option gives us an option to physically measure the amount of daylighting in a constructed building.

Option 1 – Simulation: Spatial Daylight Autonomy
- Percentage of the area that receives 300+ lux (how bright the light is) for 50%+ of the day (sDA300/50%)
- Annual sunlight exposure (visual discomfort) < 10%

Option 2 – Simulation: Illuminance Calculations
- Demonstrate that the illuminance levels will be between 300 lux and 3,000 lux from 9 a.m. and 3 p.m.

Option 3 – Measurement
- Measure the minimum and maximum illuminance levels that have been achieved in the building.

LEED Credit – Quality Views – EQc8 – 1 Point

Logically, this credit should focus on encouraging connection to the natural outdoor environment by being visually connected. Traditionally, the indoor environment was by no means considered a human's natural environment, and having more connections to the outdoors was promoted. This alleviates eye strain from staring at screens all day and maintains our natural circadian rhythm. In the past, LEED rewarded all types of views, even if it involved staring at a brick wall 3 feet away. However, LEED version 4 considers the flip side of our views. This credit exclusively applies to regularly occupied spaces, such as classrooms or offices, as compared to non-regularly occupied areas, such as mechanical rooms and bathrooms, where the least amount of time is spent.

This credit requires that 75% of all regularly occupied floor areas must have a direct line of sight for the building occupants to the outdoors through vision glazing placed between 30° and 90° above the finished floor, as well as meet two of the following four kinds of views:

1   Multiple lines of sight 90° apart.
2   Ability to see the **flora, fauna, sky, movement, or objects** 25 feet from the window.
3   Unobstructed view within three times the head height of vision glazing.
4   View factor (number of quality views with a 90° conical vision) of three or greater.

Note: Views into interior atria can count towards 30% of the requirement.

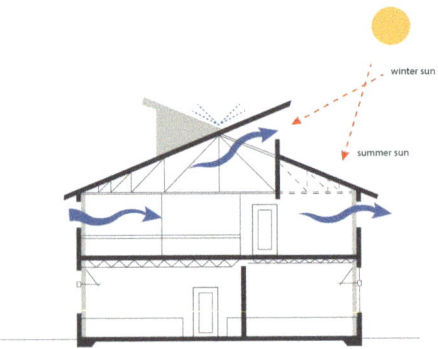

Noteworthy Standards

- ASHRAE 62.1
- ASHRAE 55
- MERV standards for filters
- CPDH and SCAQMD
- Sheet Metal and Air Conditioning National Contractors Association (SMACNA) IAQ Guidelines

Note

1 "Indoor Air Quality," EPA United States Environmental Protection Agency, Last modified September 7, 2022, https://www.epa.gov/report-environment/indoor-air-quality

10 Innovation in Design

Personally, I like to refer to the following 2 credit categories as '10 total bonus points,' as they do not logically fit in or follow the same pattern as the previous 7 categories, which account for the first 100 base points in the LEED® rating system.

As the name suggests, the Innovation in Design (ID) credit category exists to promote innovation. LEED provides credits to projects that accomplish incomparable performance in sustainability that go beyond the LEED Green Building Rating System or testing a pilot credit. Unlike other chapters, there are no prerequisites in this chapter. A project can obtain 6 points through 2 credits – Innovation (5 points) and having a LEED Accredited Professional (LEED AP+) on the project (up to 1 point).

LEED® Credit – Innovation – 1–5 Points

The ID credit category rewards projects up to 5 additional points for embracing innovative strategies and sustainable building practices to improve the performance of a building. The first credit in this category is known as innovation, and it contains three unrelated options.

DOI: 10.1201/9781003405856-10

Medal icons created by Vectors Market - Flaticon

Option 1 – Innovation (1 point)

The LEED rating system is not completely revamped or updated frequently. However, modern technology and inventions are rapidly improving and cannot be stopped from being created. The innovation credit was designed to ensure projects are rewarded for technologies and strategies that were not available or feasible when the current version of LEED was initially created. When a project completes a sustainable technique that was not listed in the previous categories but still manages to satisfy the triple bottom line, LEED offers an option to teams to create their own credit. LEED wants to reward all strategies outside of the scope of the existing rating system, which promotes a healthy environment, a prosperous economy, and a cohesive, comfortable society.

Much like a real credit, the innovation credit must possess a positive intent, quantifiable requirements, and strategies to achieve it. My favorite example of an innovation credit is using the building as a teaching tool. Perhaps you have seen a smart touch screen in LEED building lobbies that display real-time energy consumption data in addition to interactively illustrating the technologies of the LEED building. The basic notion behind this credit comes from the Prius effect. The more aware each occupant is about the operations of the building, the more likely they are to reduce their impact inside and outside the building consciously.

Alternative strategies to achieve this credit include:

- Providing an educational program covering green strategies and the environmental and human health benefits of adopting green building practices.
- Developing an organic Landscaping and Integrated Pest Management Program.
- Designing a comprehensive Waste Management and Diversion Program to divert a significant volume of waste. This integrated program can

expand to provide services to the community by allowing other buildings' users to bring in certain materials for collection and recycling.

- Formulating an environmental Green Cleaning and Housekeeping Program, which is normally only a credit in the LEED O+M category.
- Selecting low-emitting furniture and finishing.
- Utilizing high-volume fly ash in concrete to reduce its carbon footprint.

Option 2 – Pilot Credit (1 point)

Early on in the textbook, we discussed how new LEED versions are created but did not highlight who exactly tests these new credits or thresholds to assess their feasibility. LEED incentivizes project teams to single out and test drive a credit from the pilot credit library and submit feedback regarding its success or challenges. Regardless of whether the credit achieves its objective or not, the real project's feedback is the ideal solution to alter, delete, or use the new pilot credit in future LEED rating system versions. This option is achieved by:

- Accessing the USGBC's LEED Pilot Credit Library
- Registering pilot credits through the USGBC's LEED Pilot Credit Library
- USGBC member companies can submit pilot credits.
- A pilot credit proposal includes:
  - Survey feedback questions
  - Submittal Documentation
  - Identified Guest Expert

Option 3 – Additional Strategies (1–3 points)

The following option gives project teams an opportunity to earn additional points in the above two options and adds a new means of achieving an extra 1–2 points in exemplary performance. So, a project can earn up to:

A   1–3 more points through additional innovation strategies in option 1,
B   1–3 points by testing 1–3 more pilot credits in option 2, or
C   1–2 points by pursuing an exemplary performance point explained below or a combination of A, B, and C to a maximum of 3 points.

Exemplary performance is an option in some existing credits in the rating system, discussed in the previous seven credit categories. LEED wants to reward project teams with exemplary performance points for displaying an exceptional performance beyond the LEED credit requirements, commonly achieved by exceeding the highest credit threshold or, at times, doubling the credit requirements. For instance, if a LEED BD+C project achieves a 50% reduction from the ASHRAE 90.1 baseline in the LEED credit for

Optimizing energy performance, they receive 18 points. Similarly, in water, if the project achieves a 50% reduction of the EPAct in the LEED credit indoor water use reduction, they attain 6 points. But what if a project goes beyond the threshold, say, 60% more water and energy efficiency as compared to the baseline? In such an event, the exemplary performance option will contribute 2 extra points for exceeding the highest threshold.

However, this option is only made available to some credits. For example, the LEED for sensitive land protection credit awards a project for not building on sensitive farmland. In this situation of a yes or no type of credit, exemplary performance is definitely not an option.

LEED Credit for LEED Accredited Professional (1 Point)

The LEED rating system rewards project teams an additional point when at least one principal participant on their team is a LEED Accredited Professional (LEED AP+) with a specialty appropriate for the project. In most cases, the participant is the LEED or sustainability consultant on the project, but at times, they can also be other stakeholders, such as an engineer or architect. In LEED version 4, the LEED AP specialty must align with the project they are working on. For example, if the team member is a LEED AP BD+C, they are worth 1 point on a LEED BD+C project but not on a LEED for Neighbourhood Development (ND) project if they have not passed that exam yet. No matter the number of LEED APs in the project, only **1 point** can be earned under this credit. This credit essentially promotes the design integration to streamline the project certification process by involving LEED APs to educate other project team members and coordinate between them.

11 Regional Priority

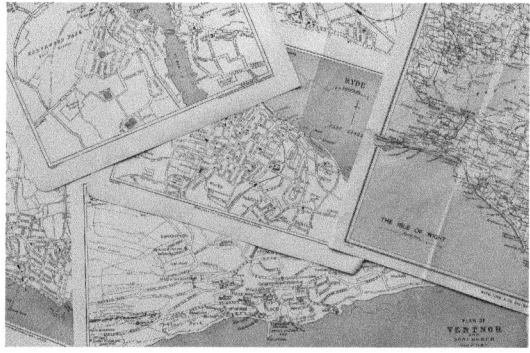

The Regional Priority (RP) category addresses geographically specific environmental priorities. For example, in areas of drought, the local green building chapter elects the six most essential credits for their region, and each of those credits is worth an extra point, provided the credit's requirements are met (up to 4 points max).

Previously we discussed a selling point of LEED®, which is the fact that it is internationally consistent, and the same rating system applies to all buildings regardless of their location. However, at times this can also hinder projects in a specific region or climate, and this reason is partially responsible for the creation of the RP category. But more importantly, the category exists to promote and stress regional-specific priorities by adding weight to certain credits.

LEED Credit for Regional Priority (1–4 Points).

The USGBC regional chapters identified different environmental regions and selected existing credits in the rating systems to be assigned as a priority for each region according to the most critical environmental issues. RP credits are listed by state and determined based on a **new GIS-based program.**

DOI: 10.1201/9781003405856-11

The RP credit zones in LEED v4 were created employing a GIS-based program that allowed for environmental issues to be empirically mapped. This process led to the emergence of RP credit zones that are based on these issues that are conducive to the environment rather than physical location. The GIS-based zones do not have to be geographically adjacent, meaning a project's RP credits can no longer be identified through ZIP codes. Instead, the project teams need to enter their physical coordinates (X, Y) to identify accurately the RP credits associated with the project.

LEED online determines the project's RP credits based on the project's (X, Y) location. Consequently, the project team **automatically** earns 1 point in addition to any points earned in the credit.

For more information about boundaries, visit www.usgbc.org/rpc.

Even though six RP credits are listed, teams can only earn up to 4 points in this category, despite complying with all six RP credits.

Example. In Arizona (X, Y), where it is extremely hot, sunny, and arid, the credits listed below are designated as RP credits.

- **High-priority site** – Curbs urban sprawl and promotes land reuse.
- **Optimize energy performance** – Lowers energy consumption due to high air conditioning requirements.
- **Renewable energy production** – Exploits the high number of sun hours per year.
- **Outdoor Water use Reduction** – Conserves water in an arid climate.
- **Building life-cycle impact reduction** – Minimizes energy consumption and waste associated with demolition.
- **Heat island reduction** – Lowers cooling requirements.

The project can earn an extra four RP points for achieving four to six of the above-mentioned credits.

Appendix I
Additional Reading

1 **The LEED® Green Associate Handbook – v4 Edition**
   https://www.usgbc.org/resources/leed-v4-green-associate-candidate-handbook

2 **Introduction and Overview – LEED BDC V4 (Excerpt of Important sections only)**
   http://leadinggreen.com/wp-content/uploads/2014/07/Introduction-and-Overview-LEED-BDC-V4.pdf

3 **Green Building Codes Background**
   http://leadinggreen.com/wp-content/uploads/2014/08/GreeningtheCodes.pdf

4 **LEED Glossary of important terms**
   http://leadinggreen.com/wp-content/uploads/2014/08/LEED-Glossary.pdf

5 **The Summary of Standards**
   https://leadinggreen.com/wp-content/uploads/2022/10/LEED-GA-REFERENCE-STANDARD-CHART-pages-2-4.pdf

6 **The Treatment by LEED® of the Environmental Impact of HVAC Refrigerants**
   https://leadinggreen.com/wp-content/uploads/2014/01/The-Treatment-by-LEED-of-the-Environmental-Impact-of-HVAC-Refrigerants.pdf

Appendix II

LEED Green Associate Exam Test Structure

Location: Prometric Center or online via proproctor tool
Length: 2 Hours + 10 minutes of instruction video with no breaks
Type: Multiple Choice
Scoring: A score of 170/200 is a sure pass, but 125/200 is the lowest possible score.

- No part points
- No points off for incorrect questions
- The more complex the question, the more points it is worth, but number of points is not stated.
- 170/200 is not equal to 85% of the questions answered correctly

Breakdown:
LEED® Process (16 Questions)
Integrative Strategies (8 Questions)
Location and Transportation (7 Questions)
Sustainable Sites (7 Questions)
Water Efficiency (9 Questions)
Energy and Atmosphere (10 Questions)
Materials and Resources (9 Questions)
Indoor Environmental Quality (8 Questions)
Project Surroundings and Public Outreach (11 Questions)
15 pretest questions (Not worth points)

Appendix III
Glossary

Alternative Compliance Paths (ACPs): Projects can comply with credits while not adhering to explicit requirement documentation in the reference guide through ACPs. If the project team believes they satisfy the credit's intent through a method outside of the existing reference guide they can submit documentation explaining credit compliance.

Basis of Design (BOD) document – developed at the end of the Design Development Phase and should include narratives and design strategies that respond to each category, goal, and requirement specified in the OPR.

Basis of Design (BOD) – the information included in the OPR such as systems descriptions, Environmental Quality (EQ) criteria, design assumptions, and applicable codes **in technical terms.**

A Regenerative project will be **Net Zero Energy =** use no more electricity than it generates onsite
- **Net Zero Carbon Footprint** – net zero carbon emissions
- **Water Balance** – uses only water received by precipitation
- **Zero Waste** – reuses, recycles, or composts all wastes

Bidding: the process of selecting the contractor to prepare for the construction phase.

Building Program/Owner's Program Requirements (OPR): It is the set of goals and requirements defined by the owner or client. This program includes the information needed for the project team to start the pre-design phase. It describes project goals, environmental vision, budget, schedule, and physical properties of a project's internal and external spaces.

LEED® refers to building program as Owner's Program Requirements (OPR).

Commissioning: process to achieve, verify, document the facilities and systems will perform as designed and installed. This ensures building quality through on-site verification.

Commissioning: The Commissioning Agent or Authority (Cx) ensures that the owner's program requirements (OPR) are included in the design process and that the building systems are installed and designed properly. LEED describes who can be the project's commissioning agent and their responsibilities through design phases. As a minimum, LEED requires the commissioning of energy systems of the project and results in reduced energy use, reduced contractor callbacks, better building documentation, and system verification.

Construction Documents (CD): Preparation of the detailed construction documents which are needed for permissions and bidding.

Credit Harmonization: Credits and prerequisites of all LEED rating systems are consistent and aligned to make it easier for project teams to switch between rating systems. Credit Harmonization promotes consistency between the many LEED Rating Systems.

Credit Synergies: When making decisions regarding a strategy that would comply with credit requirements, a project team must analyze this decision with respect to other possible interacting credits. These credit interactions can have synergies or tradeoffs between them. An example of these interactions is the relationship between storm water management credit and the water use reduction or between the day lighting and views credit and optimizing energy use.

Design Development (DD): More detailed design of the building and its energy systems.

Embodied energy is the energy consumed during the different stages of a material's life cycle and is included in its cycle assessment.

Energy Star and LEED: Energy star is developed by the EPA to allow a building owner or manager to evaluate the building's performance through free technical tools and resources. Energy Star Labeled buildings use about 35% less energy than traditional buildings. Electronics, home appliances, heating and cooling equipment can be Energy Star Labeled after they achieve certain energy efficiencies. EPA's Portfolio manager is an online tool which benchmarks your building against other and tracks consumption.

LEED, on the other hand, provides only building certification, addressing a wide range of green building features through its categories: sustainable sites, water efficiency, energy and atmosphere, materials and resources, indoor environmental quality, and innovation.

Flexible Design: Sustainable design promotes flexible design that can support future building occupancy. Team members need to look beyond the owner's current needs to include future expected needs and occupancies. LEED encourages retrofitting existing buildings because of the economic and environmental costs of a new build. Reusing buildings for a different purpose other than the first one they were built for is called **Adaptive Reuse**.

Full Time Equivalent (FTE): Many prerequisites and credits require an estimated occupancy count and LEED accomplishes this through FTE calculations. One FTE is equal to a 40-hour work week. Thus, if the sum of all 'people-hours' spent in the building over a week is 4,000, we divide this be 40 to equal an estimated 100 FTEs in this building.

Hard Costs: Costs related to construction phases like concrete, roofing, finishing materials, site work, etc. which are paid to the contractor and material suppliers.

LEED Interpretations

- Precedent setting version of Credit Interpretation Rulings
- A LEED Interpretation requires more time than a CIR because it can be applied to multiple projects and rating systems after a decision is made

LEED Pilot Credit Library: The library is compiled of a number of innovative credits which have not gone through the USGBC's drafting and balloting process for approval. Projects are encouraged to test LEED pilot credits through the Innovation in Design credit category as a means of advancing LEED.

LEED Technical Advisory Groups (TAGs): Technical Advisory Groups (*TAGs*) provide LEED technical advice; they assess and recommend technical solutions to the superior **LEED Steering Committee (LSC)** for review and approval. TAGs are responsible for providing technical advice to LEED committees and working groups to improve credits/prerequisites and support tool development. There are 6 LEED TAGs which represent the first 6 credit categories.

Occupancy: the Certificate of Occupancy has been issued and the building can be occupied.

Operations & Maintenance Program: O&M includes training the facility manager, project owner, and building occupants on how to operate the building and to optimize its performance while avoiding system degradation. The operation and maintenance program ensures that the building operates as designed and that maintenance personnel provide quality and regular maintenance to the building ensuring economic payback goals are met over time.

Owner's Project Requirements (OPR) – document is a high-level outline of the goals and requirements that are deemed by the owner to be important for the success of the project.

Project Credit Interpretation Requests/Rulings (CIRs): If a project team needs any clarification regarding any credit or prerequisite, then they may send a credit interpretation request at a cost of $220 for each request. Credit interpretation requests can be submitted by the project team at any time after the project registration. CIR requests must not exceed 600 words or 5,000 characters including spaces. No attachments, cut sheets, or drawings are allowed within a CIR request except for LEED ND where a site plan can be attached.

Credit interpretation rulings are the reviewers' responses to these requests, credit interpretation rulings constitute precedents. If a project team encounters unclear issues, they should:

 search the reference guide for help

 contact USGBC customer services for answers

 send a credit interpretation request

Schematic Design (SD): Preparation of preliminary design options and project layout.

Soft Costs: Costs outside the construction site like architectural fees, engineering fees, permits, and legal fees.

Systems Thinking Approach: Systems thinking approach is used to recognize that the change of one part of system will affect other parts of the system. It is important to understand the relationship between all parts of a system and to consider the influence of disrupting it. The goal of the systems thinking approach is to avoid designing a solution to one problem that results in another problem. For example, using triple glazed windows instead of single glazed will require a stronger structure and a smaller HVAC system to treat the space. There are two types of systems.

Open loop systems:

Open loop is the linear loop in which resources are extracted, manufactured, used, and turned into waste. It is known by Cradle to Grave open loop. Open loops are considered **unsustainable**.

Closed loop systems:

Closed loop is the circular loop in which resources are extracted, manufactured, used, and then reused or recycled. It is known by Cradle to Cradle closed loop. It is **more sustainable** and environmentally responsible.

The systems thinking approach also involves system feedback in the form of two possible loops:

Positive Loop (amplifies) – climate change melts ice → reducing ability to reflect the sun's heat using its albedo → more warming of our atmosphere.

Negative Loop (controls/regulates) – Thermostat set to 72°F in winter → room cools → thermostat tells heater to provide heat up to 72°F → Thermostat tells heater to shut off until needed.

Value Engineering: Value engineering is cutting project costs, often during construction phases affecting the actual value of the project. Many times, green technologies and features are 'value engineered' out of the design because **the most important aspect of the building is to meet the local building code.**

References

Eicholtz, Piet, Nils Kok, and John M. Quigley. "The Economics of Green Building." *UC Berkeley: Berkeley Program on Housing and Urban Policy.* September 15, 2010. Retrieved from https://escholarship.org/uc/item/3k16p2rj

EPA United States Environmental Protection Agency. "Fast Facts on Transportation Greenhouse Gas Emissions." Last modified July 14, 2022. https://www.epa.gov/greenvehicles/fast-facts-transportation-greenhouse-gas-emissions

EPA United States Environmental Protection Agency. "Indoor Air Quality." Last modified September 7, 2022. https://www.epa.gov/report-environment/indoor-air-quality

EPA United States Environmental Protection Agency. "Land, Waste, and Cleanup Topics." Last modified December 8, 2022. epa.gov/osw/conserve/rrr/imr/cdm/pubs/cd-meas.pdf.

EPA United States Environmental Protection Agency. "Learn About Sustainability." Last modified November 14, 2022. https://www.epa.gov/sustainability/learn-about-sustainability

GBCI. "About GBCI." Accessed November 1, 2022. https://www.gbci.org/about

Merriam-Webster.com Dictionary, s.v. "greenwashing." Last modified December 7, 2022. https://www.merriam-webster.com/dictionary/greenwashing

NASA. "What Is the Greenhouse Effect?" Last modified December 16, 2022. https://climate.nasa.gov/faq/19/what-is-the-greenhouse-effect/

Rosane, Olivia. "50% of U.S. Lakes and Rivers Are Too Polluted for Swimming, Fishing or Drinking." *World Economic Forum,* April 5, 2022. https://www.weforum.org/agenda/2022/04/50-of-u-s-lakes-and-rivers-are-too-polluted-for-swimming-fishing-drinking

Roser, Max. "The World's Energy Problem." *Our World in Data,* December 10, 2022. https://ourworldindata.org/worlds-energy-problem

United Nations. "68% of the World Population Projected to Live in Urban Areas by 2050, Says UN." May 16, 2018. Accessed November 1, 2022, https://www.un.org/development/desa/en/news/population/2018-revision-of-world-urbanization-prospects.html

United Nations Economic Commission for Europe. "Life Cycle Assessment of Electricity Generation Operations," 2021, https://unece.org/sites/default/files/2021-10/LCA-2.pdf

USGBC. "LEED Certification for Neighborhood Development." Accessed November 1, 2022. https://www.usgbc.org/leed/rating-systems/neighborhood-development

USGBC. "LEED Certification for New Interior Spaces." Accessed November 1, 2022. https://www.usgbc.org/leed/rating-systems/new-interiors

USGBC. "LEED Rating System." Accessed November 1, 2022. https://www.usgbc.org/leed

USGBC. "Mission and Vision." Accessed November 1, 2022. https://www.usgbc.org/about/mission-vision

Index

Note: **Bold** page numbers refer to tables; *italic* page numbers refer to figures.

acoustics 119–120
action-oriented proactive approach 76
adaptive reuse 103
air testing 125
alternative daily cover (ADC) 103
American Council for an Energy-Efficient Economy (ACEEE) 51
American Society of Heating, Refrigerating and Air-Conditioning Engineers (ASHRAE) 86, 88, 117
appliances 71, 75
artificial warming 2
ASHRAE 52.2-1999 124
ASHRAE 52.2–2007 120
ASHRAE 55-2010 125
ASHRAE 62.1-209 117
ASHRAE 62.1-2010 117, 121
ASHRAE 90.1 88, 133
ASHRAE 90.1-2010 87
asthma 115

backlight-uplight-glare (BUG) method 66
basis of design (BOD) 83
bio-based materials 109
bio-based product 108
blackwater 70
brownfield site 44
budget calculation method 123
buildable land 45
building automation system (BAS) 92
building costs *8*
building footprint 40

carbon offsets 97–98
cartridge-based units 71
Center for Resource Solution 97
chain of custody 108

Chartered Institution of Building Services Engineers (CIBSE) 117
chlorofluorocarbons (CFCs) 92–93, *93*
climate change 1, 7, 12, 24
commissioning: benefits 83; enhanced systems 85; envelope 85; mandatory systems 84; process activities 83–84; quality-oriented process 82–83; verification 82–83
conservation, water efficiency *vs.* 71
construction documents (CD) 83
construction waste **102**
Construction Waste Management Plan 100, 102
continuing education (CE) process 14–15
cooling tower water usage 78–79, *79*
Core and Shell building 15
corporate sustainability reports (CSRs) 108
cradle to cradle 100
cradle to grave stages *8*
Credential Maintenance Program (CMP) 14–15
credit compliance: abandoned/blighted building 104; building and material reuse 104; historic building 104; life-cycle impact reduction 104–105
credit rewards 73, 75, 86
credit synergies 63

daylighting 128–129
demand for energy 2, *3*
demand response 81, 95–96
demand response event 96
desalination 71
development footprint 40, 54
direct emissions 97

drip irrigation systems 74
dual flush 71
dwelling unit 27

emissions: carbon 81, 95; direct 97;
    greenhouse gases *2,* 13, 39, 47,
    81, 89; indirect 97; transportation
    systems 39, 40, 105
emittance 62
energy and atmosphere: achieve reduction
    89–90, *90*; building-level energy
    metering 91–92; CO₂ emissions
    per capita *vs.* GDP per capita *82*;
    commissioning and verification
    82–85; demand response 95–96;
    demand response programs 81;
    fossil fuels 80; greenhouse gas
    emissions and operational costs
    81; green power and carbon offsets
    97–98; LEED 80–98; minimum
    energy performance 85–88; non-
    renewable resource extraction
    and transportation 81; refrigerant
    management 92–94; refrigerant
    usage 82; renewable energy
    production 94–95
energy consumption *2,* 81, 86, 87, **87**, 91,
    92, 94, 96, 97, 128, 132
energy demand 80, 86, 89, *90, 91,* 96
energy efficiency 80, 86, 90, 134
energy metering 81, 91–92
energy modeling 81, 86, 87, **87**, 88, 95, 128
Energy Policy Act of 1992/2005 (EPAct)
    74, 75
energy production 2–3, *3,* 82, 96
energy rates *96*
energy resources 2, *3*
energy star 75
Energy Star Portfolio Manager 89
environmental product declarations (EPDs)
    106–108
Environmental Protection Agency
    (EPA) 1
environmental reduction 104
Environmental Site Assessment *55*
environmental, social and governance
    (ESG) 111
environmental tobacco smoke (ETS) 118
EPA's Target Finder 89
erosion 55
excavated soil 103
exemplary performance 133–134
extended producer responsibility (EPR) 109

extensive green roof 63
extraction reporting 108–109

floor area ratio (FAR) 45
flush out 121, 125
Forest Stewardship Council (FSC) 109
40/60 rule *20,* 20–21
Freon 93
full-time equivalents (FTEs) 71

gallons per flush (gpf) 71
gallons per minute (gpm) 71
Gerrymandering 40
Global Warming Potential (GWP) 93
graywater (greywater) 70
Green Associate Exam 71
green buildings 3; economic benefits 6,
    8; environmental benefits 6, *8*;
    evaluation 5; LCA 7; LCC 7–8, *8*;
    social benefits 6, *8*; triple bottom
    line *5*
Green Business Certification Inc. (GBCI)
    11, *11*
green-e 97
green e-certified provider 97
greenhouse gases 1, 2, 3, 39, 81
green power 97–98
Green Raters 19
green roof: credit synergies 63; intensive
    64; parking 64; solar panels *64*
green vehicles 51–52

hard costs 7
heat island effect 61–62, *61*; non-roof
    62–63; roofs 63; shading 63
Historic District 44
Home Energy Rating System Rater (HERS
    Rater) 19
household wastewater 70
hydrochlorofluorocarbons (HCFCs) 93
hydrofluorocarbons (HFC) 93
hydrology 56

indirect emissions 97
indoor air quality (IAQ): air models 121;
    assessment 124–125; breathability
    116; carbon dioxide monitoring
    121; commissioning process 116;
    construction *124*; construction
    management 123–124; enhanced
    strategies 120–121; entryway systems
    120; factors *120*; filtration 120;
    increased ventilation 121; interior

cross-contamination prevention 120; minimum performance 116–117; source control and monitoring 121; square footage 116

indoor environmental quality: daylight 128–129; ETS 118; IAQ (*see* indoor air quality (IAQ)); interior lighting 126–127; issues 116; LEED 115–122; low emitting materials 121–123; minimum acoustical performance 119–120; minimum indoor air quality performance 116–117; occupants 116; quality views 129–130; thermal comfort 125–126, *126*; ventilation *117*; ventilation rates/ contaminants 116

indoor water use reduction 74–76

infill site 44

ingredient optimization 111

innovation 132–133

innovation in design (ID): credit categories 131–132; exemplary performance 133–134; LEED AP+ 134; pilot credit 133

integrated design process (IDP) 34, *36*

integrated project delivery (IPD) 35, *36*

integrated project team *34*, 34

integrative process (IP) 35; energy systems 37; phases *37*; water systems 37

intensive green roof 63, *64*

interior lighting 126–127

International Organization for Standardization (ISO) 107

International Plumbing Code (IPC) 70

Internet of Things 92

interstitial space 113, *113*

kilowatt hour (KWH) 3

land allocation *2*

land-clearing debris 103

leadership extraction practices: bio-based materials 109; EPR 109; materials reuse 109; recycled content 109; wood products 109

Leadership in Energy and Environmental Design (LEED) 11–13; accreditation 14; bulk certifications 20; certification 13–14; challenges 7; consumer 12–13; cost 7; developer 12; environment 13; goals 3; Homes *18*, 18–19; Homes Providers 19; Interior Design and Construction 16, *17*; multiple certifications 18; Neighborhood Development 19; Operations and Maintenance 16, *17*; professionals 10; rating system 9, 12, 15–16 (*see also* rating system; U.S. Green Building Council (USGBC)); reference guide 24; registration and certification 27; stages of sustainable practices *4*; traditional practices *vs.* sustainable practices 4–5; type and size of building project *15*

LEED Accredited Professional (LEED AP+) 14, 29, 131, 134

LEED credit, location: compliance 45; diverse uses 46; high-priority site 43–44; high property density *45*; Neighborhood Development location 42, *42*; redevelopment 44; sample excerpt *47*; selecting a site 42–43; sensitive land protection 43; surrounding density and diverse uses *44*, 45

LEED credit, transportation: access to quality transit 47–48, *47*, *48*; bicycle facilities 49–50, *50*; reduced parking footprint 50–51, *50*; school specific option 49

LEED Fellow 14

LEED for Building Design and Construction (LEED BD+C) *16*

LEED for Interior Design and Construction (LEED ID+C) 16, *17*

LEED for Neighborhood Development (LEED ND): Built Project 19; credit categories 20; planning phase 19

LEED for Operations and Maintenance (O+ M) 16, *17*

LEED Green Associate 14

LEED Green Associate Exam 9, 13–14, 15–16, 86, 87, 122–123

LEED Green Building Rating System 131

LEED online *28*; calculation tool 75; owner and agent 28; Project Administrator 28

LEED Pilot Credit Library 133

LEED's Fundamental Refrigerant Management 94

LEED's water efficiency 53–79

life cycle assessment (LCA) 8, 100

life cycle cost (LCC) 7, *8*
lighting control 126
lighting quality 127
light pollution 66–67
Illuminating Engineering Society
    (IES) 66
location and transportation (LT) 39–40;
    credit category 40; emissions 40;
    green vehicles 51; LEED credit
    (*see* LEED credit, location; LEED
    credit, transportation); project
    boundary 40, *41*
location valuation factor 105

mandatory commissioned systems
    (MEP) 84
material disposal 99
material ingredient reporting 111
material reuse/recycling 100
materials and resources: building life-
    cycle impact reduction 103–105;
    construction and demolition waste
    management planning 102–103;
    credit metrics and cost calculations
    101; demolition debris 99–100;
    LEED 99–105; life cycle impacts
    100; MRcHealthCare 113–114;
    non-hazardous construction 100;
    product disclosure and optimization
    105–112; purchasing policies 100;
    storage and collection, recyclables
    101; sustainable selection 99;
    Toronto *106*; waste reduction 100–
    101; waste stream 99
material selection 99
materials reuse 109
mechanical ventilation 117
micro-misters 74
minimize potable water usage 70
minimum efficiency reporting value
    (MERV) 120, 123–124
minimum energy performance: ASHRAE
    86, 88; building projects
    102; energy consumption 86;
    prescriptive compliance 88;
    synergies and tradeoffs 86; whole
    building energy simulation 86–88
minimum program requirements (MPRs)
    25–27
mixed construction project *21*
mixed-mode ventilation 117
mixed-use sustainable communities 8
Montreal Protocol 93, *94*

MRcHealthCare: design for flexibility 113;
    furniture and medical furnishings
    113; PBT source reduction
    (healthcare) 113–114
multi-attribute optimization 107

natural refrigerants 93, *94*
Natural Resources Conservation Service
    (NRCS) 51–52
natural ventilation 117
neighbourhood development (ND) 134
net metering 95
net-zero/net-positive buildings 7
non-cartridge-based units 71
non-renewable resources 2, 81

occupancy 12, 71, 90, 123, 124
open grid pavement *60*
outdoor water use reduction 73–74, **74**
owner's program requirements (OPR) 37,
    83, 89
Ozone Depletion Potential (ODP) 93

peaker plant 96
persistent bioaccumulative toxins (PBTs)
    113–114
piping water 71
post-consumer materials 108
potable water 70
pre-consumer materials 108
process water 70
product category calculations 122
product disclosure/optimization: EPDs
    106–108; life cycle impact 105;
    location valuation factor 105;
    material ingredients 110–112;
    material ingredients/contents 105;
    sourcing of raw materials 108–109;
    transportation emissions 105
property boundary 40
protect absorptive materials 123
purchasing policies 100

rainwater management 58–60, 63, 73
rating system: AP exams focus 22–23;
    credit categories 23; development
    24–25; guidance selection 20;
    points 38, *38*; steps for creating a
    new *26*; valued impact category
    24, *25*
raw material source 108–109
recycled content 109
refrigerant management 92–94

regenerative buildings 7–8
regional priority (RP): category 136; creation
    135; environmental priorities 135;
    GIS-based program 135
renewable energy 80, 81, *84,* 95
renewable energy certificates (RECs)
    97–98, *98*
renewable energy production 94–95
runoff water *see* stormwater runoff

schematic design (SD) 37
secondhand smoke *see* environmental
    tobacco smoke (ETS)
sedimentation 55
Sheet Metal and Air Conditioning
    National Contractors Association
    (SMACNA) 123
sick building syndrome (SBS) 115
Smart Growth 41
SmartMeter *92*
soft costs 7
solar panels *64*
solar reflectance (SR) 62
solar reflectance index (SRI) 62
sound transmission class (STC) 119
sprinkler systems 73
stormwater management 58
storm water management plan (SWP) 58
stormwater runoff 70
supply chain optimization 111–112, *111*
surface area 58–59, 78, 104
sustainability *1,* 1, 11, 53, 76, 86, 131, 134
sustainable building design 91
sustainable constructions 3; *see also* green
    buildings
sustainable design 128
Sustainable Forestry Initiative (SFI) 109
sustainable sites (SS) 53–54, *54;*
    construction activity pollution
    prevention 54–55; Environmental
    Site Assessment *55–56;* erosion 55;
    green infrastructure 59; healthcare
    facilities 67–68; heat island effect
    61–62, *61;* joint-use facilities 67;
    lighting zone *66,* 65–66; light
    pollution 66–67; light pollution
    reduction 65, *64–65;* low impact
    development 59; open space 57–58;
    rainwater management 58–60,
    60; sedimentation 54–55; Site
    Assessment 56; site development
    57; site master plan 67; tenants 67,
    68; vegetated roofs 58

thermal comfort 90, 125–126, *127*
thermal difference, developed *vs.*
    underdeveloped areas *61*
thermograph, developed area *62*
tradable renewable certificates (TRCs) 97
transportation emissions 39, 105
triple bottom line 5, *5*

Uniform Plumbing Code (UPC) 70
United Nations Environment Program 69
urban planning theory 41
urban system 3
U.S. Environmental Protection Agency
    (EPA) 100
U.S. Green Building Council (USGBC) 9;
    application process outline 30–31;
    certification process 31; committed
    10; Credit Forms and Calculators
    28; educational opportunities 10;
    eligibility 29; greenbuild 10; green
    building resources 10; legitimate
    10; marketable 10; mission 10;
    networking 10; project checklist 29,
    30; review process 31; vision 10

value engineering 35
vegetated roofs 54, 58
ventilation systems 117
volatile organic compounds (VOC)
    121–122

waste *2;* construction and demolition
    management 102–103; diversion
    rates 102, *103;* landfill-bound
    matter 101; reduction 100; stream
    99; *see also* waste management
    strategies
waste management strategies: recovery
    101; recycling 101; reuse 101;
    source reduction 100
waste-to-energy systems 103
water distribution systems 76
water efficiency: blackwater 70; *vs.*
    conservation 71; cooling tower
    water usage 78–79, *79;* dual flush
    71; full-time equivalents (FTEs) 71;
    gallons per flush (gpf) 71; gallons
    per minute (gpm) 71; graywater
    (greywater) 70; indoor water use
    reduction 74–76; LEED 70–79;
    outdoor/indoor water reduction 70;
    outdoor water use reduction 73–74,
    74; potable water 70–71; process

water 70; renewable resources 69; research 69; stormwater runoff 70; waterless urinals 71; water metering 76–78; water meters 70; WaterSense 71
water-efficient plumbing fixtures 75–76
waterless urinals 71

water metering 76–78
WaterSense 71
watersense labeled products *72*
water-stressed conditions 69
water use reduction strategies 75–76
weather conditions 2
wood products 109